GW00703026

To Anthony Lawrence

— hope you enjoy the book

Edward Whund

May 25, 2005

FIELD MANUAL FOR

OIL AND GAS CONSTRUCTION

CONTRACTS

MANAGEMENT

Edward Whitticks

Gulf Publishing Company
Houston, London, Paris, Zurich, Tokyo

FIELD MANUAL FOR
OIL AND GAS CONSTRUCTION

CONTRACTS
MANAGEMENT

Edward Whitticks

Copyright © 1994 by Gulf Publishing Company, Houston, Texas.
All rights reserved. Printed in the United States of America. This
book, or parts thereof, may not be reproduced in any form without
permission of the publisher.

Gulf Publishing Company
Book Division
P.O. Box 2608, Houston, Texas 77252-2608

10 9 8 7 6 5 4 3 2 1

Library of Congress Cataloging-in-Publication Data

Whitticks, Edward.
    Field manual for oil and gas construction contracts
management / Edward Whitticks.
        p.   cm.
    Includes index.
    ISBN 0-88415-178-6
    1. Petroleum law and legislation.   2. Construction contracts.
3. Contracts, Letting of.   I. Title.
K3915.4.W55   1993
343'.0772—dc 20
[342.3772]   93-23130
                                                              CIP

Printed on Acid Free Paper (∞)

Contents

Preface

This book has been written for major civil and engineering contractors and clients involved in large construction projects, not only in the petroleum and petrochemical industry, but in any venture where contract terms are of paramount importance and where construction claims are encountered. The book is primarily a study of contractor claims settled out of court. These case histories have never before been published.

For the client's contracts engineer, the book functions as a checklist of all the administrative actions that must be carried out from the time of contract award to closeout. Included are samples of the forms needed for these purposes, as well as suggested formats for change orders, contract amendments, and the like, and guidelines on when and how to use them.

In this respect, the *Field Manual for Oil and Gas Construction Contracts Management* may be used as a textbook for engineering and contract management students. It may be useful also as a basis for seminars developed for client and contractor project managers and engineers.

Edward Whitticks

CHAPTER ONE

Client Policy

INTRODUCTION

The oil, gas, and petrochemical industry, in the main, instructs and imposes a duty upon its employees to observe certain policies regarding the contracting process and to maintain a high standard of ethics and fairness. All contracts are issued in writing and are intended to be executed prior to committing work to a contractor or a contractor starting work. Adequate safeguards and controls in terms of procedures, documentation, files, reviews, and approvals are expected to be established and maintained for all contracting activities.

INVITATIONS TO BID

Open bidding, that is, an open invitation to contractors to tender for work, is not a usual feature of oil company practice. Advertising in technical magazines and other media is more properly confined to the building construction industry and to work for governmental departments. Work in refineries, gas plants, and on pipelines is for the most part highly specialized, and only a limited number of contractors have the resources and the work forces to undertake such ventures. In an open tender situation, however, this consideration would not discourage a large number of contractors from having a shot at it. Invitations, therefore, are restricted to those organizations who are regarded by the client as capable of handling the work.

LUMP SUM CONTRACTING PREFERRED

Lump sum contracting is normally the preferred contract form when it can be properly used. Competitive bidding is used to select the technically and financially qualified contractor whose bid represents the least overall cost to the client. Negotiated contracting is only used in exceptional circumstances involving exceptional skills or equipment which would otherwise be unobtainable.

Lump sum, unit prices, day rates, and reimbursable cost fixed fees which are quoted under competitive conditions for defined work are not normally negotiated. There may be exceptions to this policy if:

- The contractor selected is the only one capable of performing the work within the project schedule
- The selection of any other contractor would result in substantial additional costs to the client
- The contractor selected possesses special expertise or other experience which would cause its work performance to be superior to that of other potential bidders

REIMBURSABLE COST CONTRACTS WHERE NECESSARY

Reimbursable cost contracts with fixed or percentage fees for "hands on" work are definitely the last choice for clients and will be used only where there is little or no engineering accomplished in the time available or in any other situation which makes a "cost plus" agreement unavoidable. Oil company contract managers will explore every possibility before entering into such an arrangement. In the late 1970s, when the oil industry was still enjoying a major boom, this policy was partially relaxed because many contractors were reluctant to accept any other type of contract unless the risk was minimal, but in times of recession the shoe is on the other foot and lump sum is the order of the day.

Compensation to the contractor under reimbursable cost contracts generally consists of three elements:

- Reimbursable costs
- Fixed rates
- Fee

The first two elements are intended to cover the contractor's costs of performing the work under the contract. The third element is intended to cover the contractor's general costs of doing business and his profit related to the work. The term "reimbursable costs" means those costs of performing the work under the contract for which the contractor receives *direct* reimbursement as opposed to *indirect* reimbursement through fixed rates.

Reimbursable costs include only those allowed according to the terms of the contract and they must have been incurred after contract award and be net costs actually incurred and paid. They exclude the costs of materials, services, and other items which are provided for in the fixed rates. They also exclude the contractor's general costs of doing business and profit which are provided for in the fee.

Fixed rates are arrangements to pay the contractor for certain costs incurred in connection with the work under the contract. They are normally used for most of the contractor's office costs such as payroll burden for reimbursable personnel, departmental overhead, computer costs, and reproduction. They may also be used for some other costs such as construction equipment.

Fixed rates should represent the contractor's best estimates of fixed unit prices permitting him to recover costs incurred and paid in connection with work under the contract. They should not be designed as profit generators; the contractor's total planned profit is included in the fee. Some clients, as a general policy, may allow a reasonable return to the contractor on investment in construction equipment dedicated to the project if the cost of capital is not recovered in the fee or otherwise.

Fixed rates, when agreed at award and documented in the contract, become fixed unit prices and are not normally subject to revision during the course of the work.

The fee provides for the contractor's recovery of general costs of doing business and includes:

- General and corporate overhead costs
- Income, profit and taxes
- Interest on capital
- Research, development and bidding costs
- Business risks

The fee also provides for the contractor's total planned profit relating to performance of the contract work.

A fixed lump sum fee is normally used when the contractor's work may be adequately defined at the time of bidding; if this is not possible, fixed rate fees applied to reimbursable manhours will be used and are applied only to hours actually worked.

The client, as a policy, may include in the contract provisions for increasing or decreasing the lump sum fee if authorized changes affect the amount of contractor's services required to complete the work.

APPLICATIONS, ADVANTAGES, AND DISADVANTAGES

Lump Sum Applications

The lump sum or hard money contract policy may be applied when:

1. Detailed design engineering, drawings, specifications, and a defined project scope of work are complete or nearly so.
2. It is possible to find contractors willing to compete for work involving a proprietary process (i.e., in a petrochemical plant) or special equipment (e.g., definable marine work).

Lump Sum Advantages

- Client is able to define the work and prepare drawings and specifications before going out to bid.
- In view of (1) above, the client may also prepare an "in house" fair price estimate for bid comparison and know in advance his budget commitment.
- Close competition between bidders may be realized.
- Contractor has an incentive to adhere to schedule and budget.

Lump Sum Disadvantages

- There may be a long period between the decision to issue the requests for quotation from bidders and the actual contract award.
- A low bid contractor may try to recover losses through excessive variation order proposals and claims.

Reimbursable Cost Applications

The reimbursable cost or cost plus policy may be applied when:

1. The project scope of work cannot be defined in detail at the scheduled time of commencement and the engineering design, drawings, and specifications are not completely available.
2. A very tight schedule is essential.
3. The bid package period must be kept to a minimum.

4. Contractors are not prepared to accept lump sum contracts for reasons that may include work in remote areas; in countries of volatile politics and unstable government; or in highly technical and new projects with the possibility of frequent design changes.
5. Expensive and sophisticated equipment is involved such as deep sea pipelaying barges and the accompanying marine fleet.

Reimbursable Cost Advantages

- Client controls costs through audit and cost compliance.
- Client has closer participation in the project by virtue of management control.
- Contractors' profit is protected by his fee and therefore he does not need to cut corners or make cost savings to the detriment of the project.
- It may be possible to convert all or part of the reimbursable cost contract to lump sum when the scope of work becomes fully defined and all drawings and specifications are complete. This conversion would have to be included as an option in the original agreement.

Reimbursable Cost Disadvantages

- Contractor has no monetary incentive to keep client's costs to a minimum or to adhere to schedule.
- Client has no guarantee of final cost of the project.
- The cost in time and staff of checking and auditing invoices for materials and contractor's personnel is more expensive than the hard money contract conditions.
- Compensation for contractor's errors and omissions is difficult to achieve.

THE PROVISIONAL CONTRACT PRICE AGREEMENT

Given circumstances where a lump sum contract would be unsuitable and a cost reimbursable arrangement would be unacceptable, there is a device which would suit most requirements and would result in a fair deal to client and contractor. This is the Provisional Contract Price Agreement and it works as follows:

1. Bidders will indicate their Fixed Costs at cost as described in the following example. Because of the possibility of front end loading with finance charges (in situations where the successful contractor would be obliged to lay out a considerable sum of money before his first progress invoice is submitted), a proportion of the Fixed Costs is paid at the beginning of the contract and in monthly increments thereafter.

2. Quantities relating to each unit price line item are estimated (by the client) and bidders will quote on these, including profit. The total, when added to the Fixed Costs is known as the Provisional Contract Price.

3. Substantial additions or deletions to the work over the life of the contract would actuate a sliding scale which allows for a reduction in the total unit price and an increase in the Fixed Costs or vice versa.

4. The logic behind this arrangement is simple and fair to client and contractor and is based upon the premise that the award to the contractor of extra work of the same nature and using only limited extra labor and equipment should result in lower unit price cost but possibly increased fixed costs. For example, in a contract involving the installation of one mile of pipe in a trench, the client instructs the contractor to lay an extra five hundred yards of pipe and excavate accordingly. The contractor will use approximately the same amount of men and equipment and it would be reasonable to expect that the contract unit rate should be reduced. It would also be reasonable to allow the contractor more money for his fixed costs in view of the extra time involved in keeping his organization mobilized, etc.

5. In the P.C.P. arrangement the bidders are asked to quote on a sliding scale the relevant increases or decreases in prices for more or less work awarded during the progress of the contract. At the end of the contract, the final amount to be paid or deducted is calculated by application of the sliding scale.

Fixed Costs	$
Mobilization	1,200,000
Camp construction	1,500,000
Camp operation	2,000,000
Transportation	300,000

Fixed Costs	$
Safety program	50,000
General administration expenses	500,000
Demobilization	750,000
Total Fixed Costs	6,300,000

Tabulation (from Bill of Quantity)

Site preparation	1,500,000
Civils	6,300,000
Mechanical	25,500,000
Electrical	10,000,000
Instrumentation	5,500,000
Insulation	1,200,000
Painting	750,000
	50,750,000
Fixed Costs	6,300,000
Provisional Contract Price	57,050,000

Sliding Scale

Increase work by:	Decrease Unit Cost	Increase Fixed Costs
10%		
20%		
30%		
40%		
50%		
Decrease work by:	Increase Unit Cost	Decrease Fixed Costs
10%		
20%		
30%		
40%		
50%		

Policy on Claims

In times of recession, when oil companies are in a strong position regarding conditions of contract, they are still a comparatively soft touch when it comes to acceptance of contractors' claims. Performance claims, when they arise, are usually investigated promptly and resolved in a reasonable manner and most clients' project procedures will reflect this policy of fair play.

Quality and "Intent" of Contracts

In spite of the client's employees' desire to adhere to these principles in the field, they are often governed by the quality of the contracts they administer. As explained later in these pages, the field contracts engineer is not always the one who put the contract together in the first instance and therefore he is obliged, when disputes arise, to seek for the intent of the contract when it was compiled.

Some lawyers will hold that "intent" is not worth much in law and courts will only consider what is *written* in the contract, assuming that both parties chose and signed the written words thereby accurately setting out their intentions. However, this book is concerned with judgement by nonlegal personnel in the field on situations which are unlikely to appear in court. In general, oil company policy is such that, if it can be reasonably argued that the contractor did not intend to follow a certain course of action when he bid for the work, the argument will usually receive sympathetic consideration where the contract is silent on the subject or is capable of being interpreted either way. The degree of sympathy may appear to expand in circumstances where the contractor is indigenous to the country in which the client operates.

Field Administration of Contracts

A client site team operating in remote areas does not have a company lawyer in residence. In all probability, the team has never seen the company lawyer and therefore relies on the contracts engineer for immediate and local guidance. In order to give this service to his colleagues in construction, the contracts engineer must know the contract from cover to cover and he must also be aware of company policy and philosophy for the whole project.

A good example of this type of awareness was shown by a field contracts engineer who was called upon to administer a contract (which he had not written) involving the installation of a pipeline through a tunnel. The design of the line included a concrete anchor block in the sag bend in the middle of the tunnel. The prefabricated bend in the line was to be installed by the pipeline contractor but the concrete anchor block was to be cast by the tunnel contractor. However, the pipeline contract scope of work stated that the pipelay contractor was to be responsible, not only for the correct alignment of the bend but also for the casting done by the tunnel contractor. The actual wording of the contract was as follows:

"The prefabricated bend shall be pre-installed by the Contractor who will also be responsible for the casting of the anchor block by others at the tunnel low point, sufficiently in advance of the start of pipelaying to allow for the concrete to cure."

The Head Office engineers who wrote the scope did not foresee any problem with the arrangement. They intended that the responsibility of the pipeline contractor should include the casting of the block only to ensure the correct alignment of the pipe bend and that the tunnel contractor should pour the concrete as this was a civil function rather than a mechanical one. No objection was raised at the job explanation meeting or prior to the signing of the contract. In fact, it was not until a week or so before this operation was reached on the schedule that it dawned on the pipeline contractor that, strictly according to the conditions of the contract, he may have to foot the bill if the concrete was not poured to specification or otherwise failed the break tests.

The contractor sought a meeting with the client's team and asked to be released from this obligation. The immediate reaction of the construction manager was to point out that the contractor had presumably read and understood the contract before he signed it, and had therefore accepted the responsibility for the tunnel contractor's work in this respect. But the contracts engineer knew that it was not the company's policy to insist on the letter of the law if there was a reasonable chance that the intent of the contract was not to penalize the pipeline contractor for the consequence of bad work done by others. The contracts engineer managed to obtain agreement to amend that particular clause in the scope to a general agreement for the pipeline contractor to supervise the concrete pour without accepting responsibility for its failure and possible cost of rework. In the event, the operation was carried out smoothly and without concrete failure so the confrontation did not take place, but some of the client's team still thought that the contractor had been let off lightly. The contracts engineer, through past experience, knew that if something had gone wrong with the pour and a claim situation had arisen, the company would have listened sympathetically to the pipeline contractor. Thus, the company saved time, money, and a good contractor/client relationship to have settled the matter beforehand.

PLANNING POLICIES

This book is not specifically concerned with the investment analysis techniques of the owners and unless he is required to provide funding for a project, the contractor may not have more than a passing interest in the client's source of capital. However, client management groups are constantly faced with the problem of choosing between various investment alternatives, for the simple reason that the funds at their disposal are limited, and shareholders and others demand the maximum possible return on their money.

There are always at least two alternatives to consider. The proposed project being one and that of doing nothing, the other. In the latter case, it may be assumed that the present rate of return on the company's invested capital is more attractive that the return expected from the proposed project. There are, of course, some immediate expenditures which are justified on noneconomic grounds such as the rescue and reestablishment of the Kuwait oilfields and elsewhere perhaps, the imposition of governmental regulations effecting the output of a plant.

The decision to embark on a new project in the oil, gas, or petrochemical industry is almost certainly made with a view to increasing revenue throughout the anticipated life of the project. Once having made the decision, it is of paramount importance to the owner that the project is completed and off and running on schedule and inside the budget. There have been surprising exceptions to this rule and one client engaged in the construction of a liquefied petroleum gas plant was alternating between goading the contractors into greater efforts and allowing the schedule to slide backwards, depending on various influences on market conditions.

In order to examine the initial moves of an oil company about to embark on a large and expensive project, we will agree to define a "superproject" as a venture expected to cost anything over a billion dollars, with separate contracts for site preparation, civil and mechanical work, insulation, refractory, electrical, instrumentation, and painting worth from five million to five hundred million dollars each. We will consider the mechanical contract at the top end of the scale for value.

The probable order of construction in, say, a refinery extension would be:

1. Site preparation (grading, roads, paving, and the like)
2. Civils (foundations, substations, control buildings, and the like)
3. Mechanicals (steel structure, pipework, tanks, spheres, vessels, rotating equipment, and the like)
4. Insulation and refractory
5. Electrical and instrumentation
6. Painting

The client will engineer the work, probably by engaging a design engineering contractor to produce the specifications and drawings. The Head Office contracts engineers will begin the task of gathering information to make a up a bid package for each contract in the disciplines mentioned above. First, the model Articles of Agreement will be modified to suit a particular contract, and the modifications will be offered to the company lawyers for approval. Scope of work checksheets will be collected from the project engineers and several discussions and meetings will be held before the final scope emerges. A bar chart schedule will outline the work sequences within the overall time frame for accomplishing the work. Milestone dates in the schedule will be determined. The contracts engineer will advise on the options regarding the pricing of the contract and will eventually insert details of lump sum, unit rate, or other methods of payment as appropriate. This exercise may include the collection and layout of Bills of Quantity and details of payment by measurement. He will also include information on free issue materials should the company propose to adopt this route and will put all this into a bid package or request for quotation, usually in a frenzied race against time. Meanwhile, the estimating engineers will be putting together their estimate of how much the product should cost.

In the front-end planning of a superproject, most client organizations will establish a small core team well in advance of anticipated start dates. This team, made up of experienced client personnel, will prepare a project execution plan. The first issue of the plan will be brief in outlining the philosophies of how the project will be conducted. This document will be the forerunner for subsequent detailed planning. The format may be as follows:

- Overall schedule, bar chart
- Resourcing plan
- Organization chart with durations

- Contracting plan
  - Basic strategy
  - Master list of contracts
  - Dates of award
  - Approximate value
- Quality assurance plan
- Engineering schedule and plan
  - Preliminary engineering
    - Key dates
    - Definition of output required
    - Constraints given designer
  - Detailed engineering
    - Key dates
      - for procurement
      - for receipt of vendor data
      - for output (definition of output required)
- Procurement plan
  - Major items
  - Schedule
- Construction plan
  - Schedule
    - Key dates
    - Installation
    - Operation

The basic project philosophy will be defined before writing a project execution plan. One question to be addressed is, How will the project be managed?

There are a number of variations:

- By client personnel entirely
- By client with personnel support from a major contractor
- By client with consultants
- By a management contractor

A superproject could well employ several hundred specialist personnel extra to the client's normal staff level. Assuming that such a project will have a duration of two or three years and members of the project team will have varying dates of mobilization and demobilization, the oil company client will not normally employ home office and field staff on a permanent basis but will seek outside help. This can be achieved by using consultants or free-lance specialists, but is more usually accomplished by engaging a service contractor who has a sufficiently large force of trained personnel to manage the project. Such international giants as Bechtel, Fluor, Foster Wheeler, Kellogg, Parsons, and others provide staff for these ventures and, if required, will also engineer the work. This is usually done through a cost reimbursable plus fee contract.

The managing contractor will undertake to issue contracts for the work, either in the client's name or on its own account, depending on the terms of the prime contract with the client. Apart from the advantage to the client of having the managing contractor supply staff with special skills and relieving the client of the extra involvement in payroll and wage rate determinations, fringe benefits, personnel transportation, and other labor or social aspects of hire, the managing contractor also offers engineering and design skills, established procedures and systems, and specialist knowledge which may not be available within the client's organization.

In project planning, the client may consider the appointment of one or more detailed design engineering contractors (D.E.C.) in addition to the managing contractor. This has been known to lead to inefficiencies in the ability of the managing contractor to support the client. One reason is that the D.E.C. often has no contractual interest in the overall project schedule and therefore no incentive to expedite the procurement related part of its scope of work. It is worth investigating an alternative where the managing contractor is made responsible not only for the supervision of construction but also for engineering and procurement.

Integration

In every project there has to be a certain amount of integration between the client's project staff and others in the construction effort. There could not be a watertight division between, for example, the managing contractor's staff and that of the client. On the other hand, the client's staff at the lower levels could not be in a position to supervise or issue instructions to the contractor's management. Most clients work this out amicably enough with their managing contractors. The evolution of the managing contractor is comparatively recent. Before 1958, for instance, one of the world's largest oil producers had its own construction division, mostly for civil work within its own community, and purchased very little from local merchants. Onshore construction companies in the area had no work with this client although international firms were occasionally involved with refinery renovations and the like. There was a certain amount of offshore work, mostly between Brown & Root and McDermott laybarges. At that time, daily crude oil production was around one million barrels. With the increase in oil production (2.6 mbd by 1967) and the build up of pressure from the government for participation in ownership, local purchase, and construction, the company slowly gave way and encouraged local traders and contractors to seek agencies and joint ventures with western manufacturers and contractors. By the mid 1970s, production had risen to more than ten million barrels per day and the company sought to collect and process the flash gases associated with the crude oil produced. Plans were made to complete by 1982 a gas gathering project costing an initial five billion dollars. For this project and similar ventures in the country, a leading U.S. engineering-construction company was chosen to act as managing contractor. In 1980, such companies as Fluor, Parsons, Foster Wheeler, C. E. Lummus, Brown & Root, et al., had a backlog of orders for plants worth billions of dollars, and construction contracts not only in the Middle East but in Europe, South Africa, Canada, and throughout the U.S.A. Fewer than a dozen design engineering-construction companies were in a position to accept a contract for a superproject. As a result, the expertise of the few was in great demand. Those management contractors designed the plant, ordered the material and materiel through their vast procurement channels, engaged the subcontractors for each discipline, and supervised every stage of the venture. To be sure, the oil company clients appointed their own project managers and staff who signed the documents and, in theory at least, held the purse strings but there is no doubt that the managing contractors ran the show, usually on their own terms. Had this situation been allowed to continue, it is probable that the industry would have run into massive equipment delivery problems and certainly a shortage of skilled labor. However, the recession came as quickly as the boom and managing contractors were faced with hard money quotations for what was left of the industry construction program. In some cases, they were told to find the money to finance the project if they wanted to be dealt in. Some clients decided to use the once powerful managing contractors as suppliers of labor and office services, using their now half-empty offices as headquarters for the project in hand.

Some of the managing contractors found themselves in real difficulties. During the boom years, they had acquired enormous amounts of cash because they had little or no equipment of their own, relatively few assets other than their buildings and their personnel, and they financed their day-to-day business with the client's cash. Many of them started to buy up other companies with their spare money, but when the boom ended, the other firms in the group fared no better than the purchaser. Companies who would not look at a quotation request of less than $100 million in 1980 were now seriously investigating the possibilities of a hard money paving job. Of course, large construction companies can ease the situation by laying off the workforce and even top management, but the snag is that if business improves in the near future, those laid off have either retired or found employment in alternative industries and would be difficult to recover.

In the middle of these hard times came the idea of "total integration," meaning that the client will use a onetime managing contractor to supply some of the personnel but will also recruit others through agencies, cost engineering companies, and consultancies, leaving the managing contractor no longer managing but in a new role as a "hot body shop," a kind of glorified employment agency. In theory, this should work well and project management positions would be filled by those best qualified to hold them, regardless of whether the incumbents originated from the client's permanent staff, from the managing contractor's staff, or from a small consultancy. To integrate, as applied to project policy, means to combine into a whole the best of hired help regardless of its origin.

So what is wrong with total integration? From the managing contractor's viewpoint it is nothing but bad news, however much he may try to look as though he is enjoying it. He is no longer in a position to ride along on a cost plus agreement while dealing out hard money contracts to others. He may also find that he has lost his exclusive position and may have to work alongside smaller companies in one big management organization. In a fully integrated system, the client may engage a number of contractors, consultants and small free-lance companies, or even individuals, and integrate them all into the project management team. In this arrangement, it is possible for someone on the managing contractor's staff or even the client's permanent staff to find himself reporting to another contractor's supervisor. From the client's position he loses the benefit of the managing contractor's vast contractual and procedural knowledge and generally dedicated personnel. When a managing contractor completely manages a project, he always tends to move his people to another job as they are phased out of the current one. The client, on the other hand, will probably not be able to employ integrated (and therefore temporary) staff on another project in the organization, so they will be released. This means that as the end of the project draws near, individuals will lose interest in the work knowing that there will be no further employment for them. They may, however, use every device to prolong the close out.

During the construction stage, a superproject involving a hundred contracts could have more than four hundred people on the project staff. A typical personnel distribution would be:

Project Management	10
Secretaries and clerical staff	25
Finance and administration	55
Project quality assurance	10
Project controls (cost and scheduling)	10
Safety officers	5
Procurement (purchasing, expediting, and inspection)	20
Contracts administration	15
Engineering	40
Construction supervision	200
Commissioning	10
	400

It has already been acknowledged that most clients are unable to produce personnel in such numbers from their permanent staff ranks and are unwilling to recruit directly for a comparatively short duration. In a totally integrated situation, about 75% of the superproject employees would be drawn from outside sources with varying mobilization and demobilization dates.

Governmental Policies and Influences—The "Third Party"

In most oil producing countries there exists an important authority who is not normally a signatory to construction contracts but who, nevertheless, may have a massive influence on the project. This authority may be a government department charged with the responsibility to inspect and license certain proposed activities, a commission appointed to watch over the project for potential infringements regarding safety and/or the environment, or, in very large projects, a specially convened body dedicated to one project alone. The latter authority may have a voice on matters of safety, security, the environment and/or nonconformity with government regulations and laws.

Where authorities have to give approval before the next stage of design or construction may begin, both client and contractor will find that time is not always as important to members of the commission as it is to the project. Third parties do not have the same schedule urgency as the proponents. Other authorities may not actually grant approval to contractor proposals, but may instead have the right to express disapproval, which sometimes amounts to the same thing.

It will be seen, through the example in Chapter 7, Claim No. 5, that failure to keep the government authority informed may lead to a great deal of expense for both client and contractor. When the project involves refinery work, for example, with the daily necessity to obtain permits from the client for road closures, hot work, and the like, it is all too easy to forget to advise that third party of what is going on.

CHAPTER TWO

The Bid Package

PREPARATION BY THE CLIENT'S TEAM

Inconsistencies in the contract document, of course, spring from mistakes made in the preparation of the bid package, since, pre-award changes excluded, the former document should be a replica of the latter. Therefore, the bid package Scope of Work should be written with the future field contract administrator in mind. Efforts should be made to remove all vagueness, grey areas, and indeterminate expressions. No contract, for example, should contain the conjunction "etc." as in "Contractor shall supply fuel, water, gas, etc." In contractual terms, this is quite without meaning. It is preferable to say "fuel, water, gas, and each and every item to complete the work"—if such items cannot be listed. Clear directions are important. "Contractor shall supply . . . ," *not* "provision will be made," or "electricity will be supplied."

All too often, in the client's home office, the contracts engineer is obliged to put together a bid package with input from other disciplines but does not have the time or assistance from others to check the finished product for clarity of intent. We all know that after days of proofreading the same document the words tend to swim around, and although the smallest grammatical error may be caught and corrected, a large anomaly may go undetected. It is worth the extra time to pass the draft to a colleague for a "What would happen in this event?" investigation. This may seem an elementary precaution but it is not always a procedural obligation with some clients.

Errors and Omissions

Had it been possible to keep a record of every contractor claim on every superproject over the past few years, one of the most significant factors common to each would have been inadequate or omitted

18

contract language. The oil industry construction contract which is completely watertight and which describes succinctly the obligations and intentions of both parties has yet to be written. Another rarity is the invitation to bid which is totally engineered prior to release; and yet, clients and their managing contractors continue to produce these gems in the belief that construction contractors will forgive their errors and omissions and work in a spirit of happy cooperation with a minimum of change proposals. It is true, of course, that there is a honeymoon period just after contract award when this belief seems justified, but this quickly fades away to acrimony and possible divorce and payment of alimony.

Since the Articles of Agreement are preprinted and only need fine tuning for the production of a new contract, where do the mistakes and loopholes occur? Quite a few of them appear in the nontechnical parts of the Scope of Work. Clients and managing contractors have engineers on the permanent staff who specialize in every discipline of the industry. Initial work on the preparation of a new construction contract may involve engineers from all departments including civil, vessels, pipelines, insulation, electrical, instrumentation, painting, coating, fireproofing, and other disciplines, all of whom have carefully prepared checksheets to ensure that they do not miss anything in compiling the Scope of Work. The checklists are used to fully describe the work and then the information in them is transferred to the rough draft of the scope. It is during this transfer of technical data to the contracts engineer that the message to the contractor may become distorted or unclear. One possible solution is to have Engineering write the whole of the scope with the aid of a preprinted worksheet and subsequent contractual editing by the contract engineer.

WHAT TYPE OF CONTRACT?

An additional home for potential claims is in the compensation section, especially where unit rates for extra work are to be provided.

Having gathered the Articles, Scope of Work, details of the planned schedule, and drawings and specification lists to suit the proposed bid package, the contracts engineer will work on the pricing section. The first consideration is to determine the basis on which payment will be made—lump sum, unit rates, or other options. As a general rule, lump sum conditions should not be imposed unless the bulk of design engineering is complete or likely to be complete at an early stage of the proceedings. Since this is hardly ever achievable on a large undertaking, the contracts engineer should attempt to get as many unit rates as possible into the package for extra work to be introduced during the administration of the contract.

Instructions to Bidders

The company's legal department will be consulted on changes which may be necessary to the model Articles of Agreement. Either these changes will be made directly to the Articles or a separate chapter will be inserted containing all the terms and conditions particular to the new contract, including all changes, additions, or deletions to the model Articles of Agreement. With the exception of changes and addenda during the bid stage, the bid package should be a replica of the proposed contract, plus a section dealing with instructions to bidders.

This portion, not to be part of the proposed contract, will advise bidders where to send their quotations, the closing date of the bid, the date and venue of the job explanation meeting and site visit, and special instructions regarding alternative methods of construction should the bidders come up with any bright ideas to the benefit of the project. In addition, the instruction to bidders will contain as much information about the work as possible that would not be otherwise obtained by examination of the package, specifications, and drawings. This information and questions for the recipients of the bid package may be taken from a stereotyped checklist prepared in advance by the contracts engineer. This data could be worked into the quotation request as follows.

Pricing Conditions

1. The quotation shall be valid for a period of _____
   (The bidder will fill in the number of days or months over which he is prepared to keep the quotation open).
2. Firm pricing for the duration of this Contract is requested. If the bidder cannot guarantee firm prices, the Company will consider a proposal with maximum percentage of escalation clearly stated for labor, construction equipment, and materials.
3. The quoted price(s) shall include all costs to bidder for materials, labor, equipment, testing, and each and every item of expense,

fees, taxes, overhead, and profit for bidder's complete performance of the Work as set forth herein.

4. The Company may elect to issue to the successful bidder a Contract for all field erection labor and a separate Purchase Order for all materials, shop fabrication, and freight. Bidder's quotation shall include a breakdown of these components of pricing.

5. Bidders are advised that quotations for this Work are being invited from other contractors.

Exceptions

1. Any deviations from or exceptions to specifications, drawings, terms and conditions, and/or any other documents listed herein must be clearly defined and set forth in your quotation in accordance with one of the following statements, in order to be considered for award of Contract.

Certification to Either "A" or "B" Below:

A. "Our quotation is in exact accordance with the specifications, drawings, terms and conditions, and other requirements of this Invitation to Bid with NO EXCEPTIONS."

B. "Our quotation is in exact accordance with the specifications, drawings, terms and conditions, and other requirements of this Invitation to Bid with no exceptions other than those listed below."

2. Bidder's quotation shall include the manufacturer, brand, catalog number, or other trade designation when prices are quoted on goods other than those specified herein.

Pricing

(The contracts engineer will refer to his checklist and use one or more of the following arrangements):

1. Total lump sum price for all Work.
2. Unit prices for the items set forth in Section _____
3. As a separate item, the bidder's electrical power requirements and the price for providing its own electrical power.
4. As a separate item, quote fees for compliance with any applicable regulations, local or otherwise, and for obtaining all licenses and registrations.

5. As a separate item, quote all local, municipal, state, and federal sales and use taxes, excise taxes, gross receipt taxes, and all other similar taxes applicable to performance of Work.

Schedule

1. Based on award by _____ (date), your quotation must indicate, by item, the number of weeks from award date required to:
   (a) Procure materials.
   (b) Submit legible, reproducible copies of checked fabrication and shop drawings for Company approval.
   (c) Complete shop fabrication of subassemblies (if any) after drawing approval.
   (d) Complete shipment of all items (f.o.b. jobsite).
   (e) Complete field fabrication.
   (f) Complete field erection.
   (g) Complete testing for final acceptance by Company.
2. Based on (site preparation), (foundations) being completed by others to suit successful bidder's schedule, submit your tentative mobilization arrangements and schedule.
3. Based on award by _____, quote your best completion date, by item.

Subcontracting

State which portion of the Work, if any, you propose to subcontract and list the names and addresses of potential subcontractors.

When the bid package is prepared in draft form, it will be distributed for review and comment to other client departments including:

Project management	Procurement	Construction
Engineering	Law	Audit
Accounts payable	Tax	Insurance
Cost & Scheduling	Industrial relations	
Construction quality control		

The contracts engineer is very careful to put a time limit on this review so that he has a chance to meet the bid invitation schedule. He is also at pains to ensure that nil responses are recorded, knowing very well that in the event of disputes arising during the course of the contract, some departmental heads will swear that they never approved the issue of the bid package in that form.

THE JOB EXPLANATION MEETING

For the larger or more complicated invitations to tender, clients are well advised to hold a job explanation meeting for their selected bidders and either at a later date or at the same time, conduct a site visit. This is a very important meeting for both the client's team and the bidders who will undoubtedly send along the personnel designated to manage the contract should they be successful. It is important enough for the client's contracts manager to be present and to enlist all his available staff to assist even if they had no hand in the preparation of the bid package. Engineers from the relevant disciplines should also be there to answer technical questions raised from the floor.

During the job explanation meeting, it will be helpful to have a tape recorder running. Aside from the obvious reasons, there are always one or two bidders who arrive late and rather than go through the whole performance again, the organizers will save time and breath by allowing them to listen to the playback. Bidders might bring their own machines for later listening and discussion.

The job explanation meeting should be held within a reasonable period after the distribution of the bid package to allow bidders time to review the documents and come to the meeting with questions on the work to the done. If possible, a site visit should be made on the same day.

A typical job explanation meeting agenda would contain the following items:

1. Register attendances.
2. Introduction and brief explanation of the nature of the work.
3. Examination of the Scope of Work with particular references to those areas in the scope or in the specifications which are unusual or which may need special attention.
4. Highlighting passages or references in the other parts of the contract or exhibits which may require clarification.
5. Description of the location of the work.
6. Estimated start and completion dates with special reference to items on the schedule critical path.
7. Reminders regarding closing dates for bids and where to direct them.
8. Meeting opened for bidders' questions.

Note:

It is essential that all bidders respond in the same manner to requirements in the bid package which are not clear or on which the package is silent. For example, the question is raised, Will electricity be supplied by the client or shall we provide our own generators? If the client's team does not know the answer at that time, the question can be noted and letters sent out later on. If because of time considerations etc., an answer should be given there and then, it is preferable to give a positive direction one way or the other, so that, for instance, all bidders include for supply of electricity. Even if this condition is altered later, at least all bidders are thinking along the same lines when their bids are presented.

THE SITE VISIT

It may not be possible for the site visit to be conducted on the same day as the job explanation meeting, depending on the location of the site or the time required to examine it. Nevertheless it is important that all bidders make the effort to visit the proposed site. In fact, some clients will make attendance at the job walk a condition of the bid, particularly where the proposed site is in close proximity to an existing plant, as in the case of a refinery extension.

The Articles of Agreement in Part 1 of the proposed contract handed to the bidders as part of the bid package should contain clauses confirming that the successful bidder (Contractor) has thoroughly investigated and satisfied itself regarding all the conditions of the site, including access roads, water, electricity, climatic conditions, and obstacles to be encountered such as buried pipelines, drainage systems, and the like.

Operators of refineries that have been in operation forty years or more are not always certain of the exact location of abandoned pipelines. One of the first actions of an experienced contractor following contract award at such a job site is to dig a trench all the way around the perimeter to satisfy itself that there will be no expensive surprises during the life of the contract. Of course, it will not be possible to see those hazards during the site visit, but very often a quiet word with one of the old-timers at the adjacent operating site will reveal quite a few unknown factors.

In the mid 1970s, a contractor had been awarded the job of constructing a 1,500,000-bbl. floating roof tank intended for the storage of natural gasoline on partly reclaimed land in the Middle East. Before work began on the massive foundation, all the necessary tests were carried out on the site, which had been thoroughly investigated to the satisfaction of client and contractor. But not, apparently to the satisfaction of an old company retainer who had been semiretired and relegated to the position of gatekeeper at the site from terminal operator many years before. The old man insisted that there was a large diameter outfall pipe which ran directly under the foundation to the sea. It had been abandoned for many years and had long disappeared from any plot plan but it was, nevertheless, not the sort of thing on which to balance a large tank foundation. The contractor was skeptical until he was taken a little way out to sea where the outfall pipe could be seen in the gin-clear water, heading landwards and straight under the foundation.

At the site visit to a proposed petrochemical plant in Holland, the bidders turned up at the previously arranged time of 10 a.m. and spent a couple of useful hours at the site before adjourning for lunch and returning to their respective offices. One bidder, however, had noticed railway lines passing in front of the main gate to the site over which all vehicles had to cross to get into the work area. The bidder decided to forego lunch and make a few enquiries. It appeared that, at frequent intervals throughout the day and night, locomotives with long trains of goods wagons would travel to and from the docks, sometimes blocking access to the main gate for up to half an hour at a time. The client had not mentioned this activity at the job explanation meeting or the fact that the railway authorities had no intention of changing their routes to suit plant construction. It also appeared that there was no strict timetable; the trains were dispatched along the line in accordance with the loading and unloading of ships at the docks. It became apparent that, whoever bid for this job would have to take into account the possibility of late arrival of staff, work force, equipment, and materials, not to mention the irritation of the workers having to wait to get home half an hour after clocking out!

CHAPTER THREE

Bid Preparation

INVITATION TO TENDER

Whatever time is allowed for bidders to prepare quotations to meet the closing date, it never seems quite enough. In the frenzied activity of estimating, the bidder sometimes overlooks the inconsistencies and pitfalls introduced by the client, not by design but as a result of haste or carelessness in the composition of the bid package. The Invitation to Tender (bid package) will start with a brief description of the work and how it fits in with the overall pattern of the project. It will give general instructions on how to prepare and submit the tender and what information is required concerning the bidder's organization, financial ability, joint venture intentions, and the general proposed management of the contract, if awarded. The remainder of the package is, of course, the basic contract requiring only the insertion of the bidder's quotation in the compensation section. The contract is made up of the following parts, not necessarily in this order:

- Articles of Agreement
- Scope of Work
- Schedule
- Compensation
- Materials to be furnished
- Specifications
- Drawings

Most bidders are confident that they have read the bid package from cover to cover before submitting their tenders, but in reviewing the contract *proper,* it is important not only to understand what is written therein, but also to be aware of what is left unwritten.

When bid packages and contracts are written in a language that is not the first or native language of the bidder, extra care should be taken to understand the contents and to identify the potential hazards.

Articles of Agreement

As an explanation, consider a contracts engineer writing all construction contracts from the very beginning each time, without the benefit of a model agreement. Long before he had written a dozen contracts, he would discover that there were some terms and conditions that he would be repeating over and over regardless of the nature of the work. It would not be long before these "articles" would be printed out as permanent parts of the contract and used each subsequent time with only minor alterations, if any.

The Articles, therefore, are preprinted as a permanent part of each contract and some clients consider that they are set in concrete and are not to be challenged. Unfortunately, some contractors are willing to accept this view and skip over the Articles in review as being something to be accepted without question. Indeed, most of the Articles are reasonable, practical, and inserted for purposes which few bidders would contest. On occasions, however, client's lawyers will slip in some clause which could be unfair to the contractor or at best, biased in favor of the owner. Here are a few examples:

1. "Contractor shall comply with all applicable laws and regulations of all governmental and other authorities having jurisdiction, including but not limited to laws and regulations regarding labor, taxes, and safety."

The contractor should inquire who pays if these laws are revised during the life of the contract, resulting in extra taxes, etc. Probably, the "other" authorities having jurisdiction should be listed.

2. "Without prejudice to Company's other rights under the Contract, Company shall have the right to terminate the Contract for any reason and at any time by giving notice in writing to the Contractor." (The clause goes on to explain to what compensation the contractor will be entitled if the termination is not for default.)

Most Articles are silent on any suggestion regarding the contractor's right to terminate, but if the bidder accepts this, it may be worthwhile to look closely at the compensation to determine if the amounts offered when the contract is halted through no fault of the contractor are sufficient. The same applies to the suspension clause which may be actuated solely at the convenience of the client or, perhaps, by some failure to deliver free issue material on time.

3. "In the event Contractor is a subsidiary or affiliate of or is owned or controlled in part or in whole by another company, Contractor, if requested by Company, agrees to provide a letter of guarantee satisfactory in form to Company, from that company or companies, guaranteeing Contractor's full and complete performance of the Work and all the obligations of Contractor hereunder."

This clause has been dropped from most US contract Articles because its enforceability is questionable. If it is addressed at all, it should appear in the Invitation to Bid and not in the contract.

There are other clauses in various Articles which may be questioned but it is not the purpose of this work to focus on this issue. The general message to the bidder is: "Don't be afraid to challenge the Articles if you think they may be weighted against you."

Finally, it should be mentioned that since the Articles are frequently taken from a "model" contract, one or two completely irrelevant clauses may remain in a bid package. Upon their discovery, the client will usually be quite willing to alter or delete them.

Scope of Work

Where they can, clients will insist on contracts that place the responsibility of providing *everything* firmly on the contractor, with the possible exception of materials that the client, through a larger procurement system, may buy more cheaply. These are commonly known as Lump Sum contracts and the opening paragraph of the Scope of Work may start off like this:

"Except as otherwise expressly provided in this contract, the Contractor shall supply all labor, supervision, installed and consumable materials, equipment, tools, consultation, services, testing devices, storage, and each and every item of expense necessary for the supply, fabrication, erection, installation, application, handling, hauling, unloading and receiving, construction, evaluation, design engineering, testing, assembly, and production of _____, hereinafter called the Work."

This is thought to obligate the contractor to do just about everything except walk the owner's dog, but as indicated previously, no construction contract is completely safe from attack by a determined contractor. There are always grey areas where an instruction or a condition can be read in more than one context.

In the initial examination of the scope, there are a number of checks that the bidder should make to bring up with the client at the job explanation or clarification meetings. In some cases, these things can simply be noted for future reference. The following paragraphs may be helpful in this respect.

Has It Been Done Before?

Work on a superproject is usually too technically advanced to allow an open bid situation and where international contractors or joint ventures are to be considered, the client will not normally publicly advertise the proposed contract. Such work will be bid by selected contractors who will initially receive a confidential approach soliciting their interest in submitting a competitive bid on a lump sum or cost reimbursable plus fixed fee basis or permutations of same. The work may be planned for extremes of climate, difficult terrain, or may involve technical innovations. The Flexicoker plants in South America and Europe are examples of new ventures and so are pipelines in Alaska or across the Norwegian trench in the North Sea. Undoubtedly, technical "firsts" breed design changes.

The client has an obligation to disclose essential technical information about the proposed work in the bid package and to emphasize potential difficult areas at the job explanation, if he knows about them at the time. Very often, however, problems come to light only after work has commenced. In addition, the client may know about awkward requirements in the scope or in the specifications, but is not aware at the time of bid that they will cause problems and should be mentioned specifically before award of contract.

If the Contractor is Bidding for Work in a Foreign Country for the First Time, the Following Questions May Occur:

1. How much control/influence/authority do the government agencies have?
2. Is it mandatory to accept indigenous labor?
3. How much emphasis is placed on the production of Quality Assurance/Quality Control (QA/QC) manuals and work procedures? (Do they want the contractor to produce, at his own

expense, bound volumes on how to do the complete job with full illustrations or just a few pages as a guide?) What will be the cost of staff and equipment such as word processors?

4. Is there a list of government imposed official holidays? Is there any likelihood of holidays being unofficial and unscheduled? The same goes for government imposed wage increases.

Questions on the Scope in General:

1. Is the scope reasonably clear? Can some sections be interpreted in more than one way?
2. How will client and contractor inspections at all levels be conducted?
3. Are all contractual requirements plainly in the scope or the specifications and not tucked away in a Bill of Quantity or, worse still, as notes on a drawing?
4. Is there a definition of mechanical completion? (often a contentious subject!)
5. What is the rough order of magnitude of the cost of reporting to the client?
6. Is it required to set up a training scheme for local workers (welders, for example)? If so, is there payment provision for this item or is the contractor just supposed to do it and include it in his bid?
7. Is there interface with other contractors? Do they have preference? What sort of contracts do they have? Hard money contractors have a different outlook from cost reimbursable ones.
8. Is the client supplying electric power? Will he provide standby equipment in case of mains failure?
9. If an assembly or fabrication yard is allotted to the contractor, how far away is it from the job site? A distance of a mile or so not only means costing for transportation but causes a need for additional supervision, manpower, and equipment, plus lost effort in trying to coordinate the work in two separate areas.
10. For site preparation, how far must excess dirt be hauled for disposal and if additional fill is required, where is the nearest source?
11. What classifications of labor are available locally and what are the basic wage rates and overtime rates?
12. Is there a copy of the client's project procedures available? Even if it has to be a "bootlegged" copy, it will be very handy as a

guide to how the client operates behind the scenes. The claims procedure will be especially useful.

Changes to the Scope of Work

A sample of the client's change order format should be included in the bid package or otherwise available. Contractors, being familiar with scope variation procedures, may be inclined to skip over the perusal of this form but sometimes clients include words which could be questioned by the bidders.

A standard change order will include the following paragraph:

"The payment described above shall include Contractor's profit and shall be in full and final settlement of all costs, expenses, and overheads, whether direct or indirect, incurred by Contractor and arising from or related to but not limited to: (1) the extra work described in this Change Order; (2) any delay caused to the schedule or any change in the completion date of the Project resulting from said extra work; (3) the cumulative effect, if any, of said extra work in conjunction with extra work described in any other Change Orders."

Conditions (1) and (2) are acceptable, but (3) may cause problems depending on the nature of the project. Contractors may find it hazardous to sign such a change order and still preserve their rights to impacts, acceleration, and other effects they cannot foresee.

Schedule

How Much "Float Time" is on the Schedule? Is it Possible to Claim Ownership of it?

Most contracts are silent on the question of who owns the float time but often, whichever party claims and uses it first is seen to be entitled to it.

Many a contractor has successfully claimed for delay when the delay, in fact, did him little or no harm and did not even bruise the critical path on the schedule. If a contractor does not claim ownership of float time, loss of it may result in a loss of flexibility and may eventually cost money in terms of delay or acceleration costs, or liquidated damages.

If the client is able to insert a clause into the contract claiming ownership of float time that does not effect the achievement of milestone or completion dates, he has gone a long way towards the prevention of contractor claims—so the ownership of the float time is valuable to either party.

Example: The Master Schedule indicates that the client will deliver free issue materials in Week 8 for incorporation into the permanent works by Week 12 (not critical) with overall completion by Week 20.

The client does not deliver the material until Week 10, which delays the contractor slightly but does not stop him from working in other areas of the job, or put him in standby, and still enables him to complete the job by Week 20. All the contractor has lost is the float time between Week 8 and (say) Week 14. On the face of it, no harm has been done but if the contractor had claimed ownership of that float time, he would have been in a firm position to claim.

Some clients are sensitive to this possibility and slide into the contract this innocuous looking clause:

"It is acknowledged and agreed that actual delays in activities which, according to the base line schedule, do not affect any milestone or completion dates shown by the critical path in the schedule, do not have any effect on the contract completion date or dates and therefore will not be a basis for a change therein."

The client will require a detailed work schedule showing the time to complete the work. How detailed it is may vary from client to client but may involve the employment of a logic network diagramming technique and a computer program to identify the critical/subcritical paths that establish its scheduled completion date. The network will be expected to highlight all major activities and restraints along the critical and subcritical paths.

Compensation

Assuming a bid for a hard money contract is in preparation and there are no special arrangements for advance mobilization payments, the contractor should try to get as much of his fixed costs paid at the front end as he possibly can. Even with lump sum contracts, the bid package will normally require a breakdown of the lump sum price. Depending on the methods adopted by the client, the breakdown can range from a simple division of costs for mobilization, construction, and demobilization to a complete description of the job, broken down into detailed work packages.

Regardless of how much the contractor considers demobilization will cost, it is always preferable to load the mobilization portion with as much as it can stand to get the cash flow situation off to a good start.

Unit rates for changes, where called for, should be quoted as high as possible without becoming too uncompetitive and if they are not called for, none should be volunteered.

Bidders will be asked to state the period over which their quotations will be valid. This is expected to be until approximately 30 days after the closing date of the bid.

In a proposed lump sum bid, a firm price may be requested for the duration of the contract but if the bidder cannot guarantee firm prices, the client may consider a proposal involving a maximum percentage of escalation for labor, equipment, and materials.

Materials

1. If client free issue material is involved, will the material certification certificates be readily available? Will the client commit himself in writing on this point?
2. Does the client import material duty free? Would this also extend to contractor supplied items?
3. If the contract says something to the effect that "All materials and equipment furnished by Company shall be unloaded and received by Contractor," *where* will unloading take place? (It could be at the docks, ten miles away!)
4. Where free issue steel products are provided, particularly line pipe, it is essential to know how many manufacturers are involved. Pipe from more than one mill may mean that considerably more welding procedures must be produced before work starts. If this situation is not revealed at the bid stage, one side or the other will be put to much extra expense after contract award.
5. If the contractor has to supply some or all of the materials, will the client instruct on which product to buy, or merely give the specification and allow the contractor a choice? Bidders should beware of clauses which lay down the name of the manufacturer of contractor supplied products but also hold the contractor responsible for the performance. Paint is a good example. The bid package may expect the contractor to buy paint from manufacturer X but on application it may not give the required results. Who pays for the remedial work? It is virtually impossible for

the contractor to prove that he applied the paint strictly according to the manufacturer's system instructions regarding temperature, relative humidity, etc. It is infinitely preferable for the bid package to provide the required specifications of the paint (or any such material) only and allow the successful contractor to purchase to these specifications.

Specifications and Drawings

1. Who engineered the job? How many design engineering companies were involved?
2. Do the contract specifications have precedence over international standard ones (such as API 1104)?
3. How will revisions be signalled to the contractor for "Approved for Construction" drawings?
4. What technical data and drawings are required from the bidder to allow evaluation of his technical ability to accomplish the work?

"As Built" Documentation

The certified "Approved for Construction" drawings issued to the contractor may be revised from time to time by the client through engineering design changes, but the finished product may not be exactly as the drawings dictate, although still within the scope and acceptable to the client.

The contractor, on reaching job completion, will be required to mark up each drawing to reflect the completed state. Together with these marked up drawings, he will be instructed to assemble other documentation that, when catalogued, will represent the work as it is finally built.

The physical size of the "as built" documentation will depend upon the client's requirements and has been known to fill one or two large shipping containers. Unfortunately, bid packages seldom give much guidance in this direction and at the end of the job, the contractor is faced with the considerable expense of collecting data and printing copies of practically every document that has passed through his hands. It is as well, therefore, to prepare a list of probable requirements which may include but may not be limited to the following:

- Description of change orders issued
- Exemption reports
- Agreed outstanding work list
- Signed statement by contractor certifying correct usage of materials

- Certificates showing suitability of all contractor furnished materials
- Copies of certificates for all client furnished materials
- List of "as built" drawings
- Marked-up plans and drawings showing location of all identified materials and location of tests (such as weld radiography and the like)
- Welding procedure qualification reports and welding procedures
- Welder qualification certificates
- Weld examination reports
- Non Destructive Testing procedures, qualification, and calibration reports
- N.D.T. reports
- Index of radiographs
- Mill test certificates for all installed materials
- Heat treatment records and charts
- Dimensional control reports
- Tests and Commissioning reports (hydrostatic, instrumentation, electrical, and architectural)
- Certificate of Acceptance
- All required certificates from authorizing authorities
- Any other documents relevant to certification and acceptance of the work
- Vendor drawings, manuals, and operating instructions
- Operation and maintenance manuals for contractor furnished equipment
- Test instrument calibration and test reports

BID PREPARATION IN GENERAL

Alternative Proposals

Most clients insist that quotations are prepared in strict accordance with the specifications, terms, and conditions set forth in the bid package. Failure to conform with this requirement usually provides cause for disqualification of the bid. However, bidders are not discouraged from the submission of an alternative proposal or a proposal with qualifications provided such proposals are made *in addition* to the bid prepared in accordance with the specifications.

In certain situations, bidders are positively encouraged to develop and propose alternatives and exceptions to the scope of work that will reduce the contract price, and/or improve the schedule or operating economy with no sacrifice in safety or operability. Bidders must be careful, when submitting such alternative proposals, that the only deviations proposed are listed in separate attachments to the bid response. The client will not accept responsibility for the discovery or identification of alternative methods which are mentioned in the main proposal, and should the bidder become the successful contractor, the only deviations recognized will be those written into the contract.

Subcontractors

If the bidder proposes to associate with another company for performing any portion of the work, it is usual for that company to be relegated to a subcontractor role and it would not be party to the contract. Of course, clients prefer that the work is accomplished without the use of subcontractors but if this is patently a necessity, the following data must be submitted to the client for approval:

1. Specific areas of the work to be subcontracted
2. Names and addresses of subcontractors proposed
3. Subcontractors' financial references
4. Subcontractors' experience in similar work
5. Total effect on the bid proposal should all or any of the subcontractors not be approved

CHAPTER FOUR

Evaluation

RECEPTION AND EXAMINATION OF THE BIDS

Assuming that the contracts engineer has met the deadline, all prospective tenderers will receive their bid packages and will start work on producing a quotation. The client will assemble its bid evaluation team and will open the bids on or after the closing date. Then begins the technical and commercial examination of the bids.

The period of bid evaluation is a potentially dangerous time in terms of bid security and confidentiality. The team leader, who will probably be the client's contracts manager, will arrange for the commercial and technical teams to occupy separate rooms complete with lockable file cabinets. The commercial team room, particularly, should be kept locked at the end of each day's evaluation session and at other times when the room is unoccupied. The commercial team room will be out-of-bounds to everyone not on the team, including the technical squad. Members of the commercial team, however, will require access to the technical room for clarification discussions with the engineers.

Communication with Bidders

Since the commercial team will be made up of contracts and cost engineers, communications to and from bidders by telex and telephone will be directed to the contracts personnel, who will relay queries and messages to the technical team, if necessary. It is important to discourage bidders from approaching the technical team members directly.

Checking the Bids

Most clients have a formula for technical evaluation in which each bidder is assessed on its proposed:

- Quality assurance/quality control
- Scheduling and reporting

37

- Material control
- Construction facilities
- Safety and security
- Management personnel
- Execution plan
- Experience

Of these, the execution plan, the construction facilities, and management personnel may be considered the most important and will be weighted accordingly.

The commercial team's first job will be to check the arithmetic, and it is surprising how often this is found to be in error. Included in the commercial evaluation will be the following considerations:

1. Bottom-line figures of *actual* price based on the tender document and subsequent adjustments made by the bidder.
2. Evaluated prices for outstanding items.
3. Effect of difference in cost of export financing between bidders.
4. Comparative cost of manning site team (including subcontractors).
5. Comparative effect of milestone payments.
6. Comparative effect of an increase in the scope of work. (For 10%, 20%, and 30% using the bidders rates.)

THE BID CLARIFICATION MEETING

Throughout the evaluation period, separate clarification meetings will be held with each bidder (possibly only brief meetings with those who are not seriously in contention). During these sessions, bidders will be questioned on the parts of their offer that need further explanation and important technical and commercial revelations will be confirmed in writing.

The atmosphere at clarification meetings is, by necessity, informal. No official minutes should be taken at these sessions as there is a need for relaxed discussions on contractors' modus operandi and for both sides to get a better view of the bid invitation and the response, which may not be completely apparent in the written words. It would be unreasonable to pin down the contractor to any firm statement made at the clarification meeting—hence the absence of minutes—but when an important matter arises that would affect the conditions of the contractor's offer, the client should request a separate response in writing or by telex to clarify the point raised. This written answer will be considered by the client as binding on the contractor. For example, if the contractor's bid includes for fabrication of pipework but does not specifically state that the prices include painting, the client may ask for written clarification on this score. If, later on, after contract award, the contractor produces a change proposal for painting that somehow missed the scope of work, the client would be correct in referring back to the pre-award confirmation to show that this work was in the bid.

Should a dispute arise in the form of a claim during the course of the work, the clarification meeting notes and faxes are of limited legal value because the governing document is the contract itself. Nevertheless, a contractor's claim has often been withdrawn on production of notes from the clarification meetings indicating that the bidder agreed to take this or that particular course of action or that the intent of the meeting was contrary to the contractor's current claim.

Clarification meetings also offer the engineers an opportunity to discuss the technical aspects of the proposed work. They can find out how the contractor intends to tackle certain parts, how many manhours will be expended on certain tasks, and generally question the contractor on his operational methods and resources.

There may be more than one clarification meeting held with shortlisted bidders, and in the period between the first meetings and contract award it is quite likely that several addenda will be added to the package. This usually means a flurry of telexes with price adjustments throughout the bid evaluation period.

Changes and Amendments During Evaluation

It is not unusual for changes and amendments to creep into the bid package during the evaluation period, sometimes due to alternative suggestions or offers from the bidders but mostly because of second thoughts from the engineers.

Reaching the Bottom Line

Finally, the bottom line will be reached and the evaluation team will be ready to make their recommendation to client management. To reach the bottom line, the commercial section of the evaluation team will produce a spreadsheet which will incorporate all effects on the quotations brought about by bidders' qualifications to the bid package, escalation, if any, the effects of changes not known at the time of bid but now included, the effect on the contract price by the relative time to be spent on dayrates as opposed to the lump sum portion, and the effects of various financial arrangements. Finally, the team will quantify commercial and technical qualifications and evaluate the cost to the client contained in the proposed schedules.

In the following sample spreadsheet, Bidder A, who may have started off in Row A.1 with the lowest figure, could end up the highest in Row E.1 by the time the commercial section has evaluated qualifications to the bid and quantified other considerations. Having reached an evaluated contract price in the total of A, B, C, and D, the team will then apply the effect of percentages of extra work using the bidders' unit rates for variations. This result may also make a significant difference to the final recommendation. The team's estimate of the amount of variation, to a large extent, will be based on the status of engineering work completed at the time of evaluation.

LOW BIDDING

Occasionally, the lowest bidder will be so far below the client's estimate that he could be awarded the contract only in the certain knowledge that he will lose money in the process. Some clients have strict rules to apply to this kind of situation and they will not accept such a bid for the following reasons:

1. When the "successful" contractor discovers that he is losing money on the job (if he had not anticipated it at the time of bid), he will attempt to make up the difference by change proposals and claims.
2. If another job on another project should come along during the contract, most of his attention and probably his best personnel will be transferred in that direction.

Clients who are uneasy with this situation will call in the bidder and without disclosing the amount of the client's estimate, will advise him to withdraw. Often the bidder will reply that he is aware of his low offer but has men and plant lying idle and is prepared to take on the job to keep them in work in spite of the prospect of little or no profit. This approach should also be rejected for the same reasons as given above. If *all* the bidders on the shortlist are below the company estimate, the wise client will rebid the whole job, which, in most cases, means calling in the bidders and advising them of the situation and

Sample Spreadsheet: Bid Evaluation

		Bid A	Bid B	Bid C	Bid D
A.1	Lump sum				
A.2	Escalation				
A.3	Lump sum escalated				
A.4	Dayrate work				
A.5	Escalation				
A.6	Dayrate work escalated				
A.7	Known changes				
A.8	Escalation				
A.9	Known changes escalated				
A.10	Subtotal A				
B.1	Correction to time on dayrates				
B.2	Correction for qualification to scope				
B.3	Escalation on above				
B.4	Sum: time/scope corrections escalated				
C.1	Corrections for difference in payment schedules				
C.2	Corrections for difference in financial arrangements				
C.3	Sum: financial corrections				
	Sum: total A, B and C				
D.1	Commercial qualifications				
D.2	Technical qualifications				
D.3	Schedule risks quantified				
D.4	Legal qualifications				
	Sum: total A, B, C and D				
E.1	10% extra work on unit rates				
E.2	20% extra work on unit rates				
E.3	30% extra work on unit rates				

perhaps pointing out common factors in the bid where they could have gone wrong. However, before this step is taken, it is essential to call back the company estimators and tell them to take another look at their estimate. If this is done and they have shaved off as much as they possibly can and the bidders are still below, then most clients will consider that there is no alternative but to rebid. Unhappily, not all clients take this view, believing that contractors, particularly the international ones, are big boys and should be allowed to lose their own money in their own way. In a recent inquiry for the mechanical construction portion of a petrochemical plant, all the bidders were below the company estimate, three of them at nearly half the client figure, but the award was made by management against the commercial evaluation team's advice.

The team was made up of three experienced engineers who were, as it happened, not the client's permanent staff but were seconded into an integrated project. When they discovered that all the bidders were low, the engineers recalled the company's estimators to confirm the validity of the original estimate. A few errors were discovered and corrected but the estimate was still twice the amount of most of the bids. The evaluation team, by this time, was in a position to pinpoint all the low areas in each bid and to indicate exactly where each bidder had underestimated. The team leader duly reported to the client's senior management and suggested that all five bidders should be called in, told that they were well under the company estimate without revealing what that estimate was, given the areas in which they were below, and told to rebid within two weeks. Management appeared to be horrified at the suggestion and insisted that the bids should stand as presented. The award was made in accordance with these instructions and the "successful" contractor mobilized. It is possible that some of the bidders intended to bid low and recoup on claims and change orders but certainly the one who was awarded the job soon woke up to the fact that he was in a loss-making situation right from the start. Within weeks, the contractor brought in extra help whose sole purpose was to manufacture change proposals and claims.

Every flaw in the scope of work, overlooked in the bid stage was examined for exploitation and extra work rates not already in the contract were inflated as far as possible in negotiation. Two hundred change orders and ten claims later, the company's original fair price estimate was passed and a contract price paid in excess of budget for a job not particularly well done.

A contractor who finds himself in a similar position, that is, either knowingly or in error bidding dangerously below a fair price estimate, will have to take steps to recoup his losses once he wakes up to the situation. He can attempt to do this a number of ways:

1. Try to renegotiate the price with the client
2. Extra work through change orders
3. Call in the claims experts

If renegotiation fails and he has rejected the idea of giving up altogether, the contractor will turn to change orders and claims.

BIDS ABOVE THE COMPANY ESTIMATE

The occurrence of one or some, or even all of the bids exceeding the company estimate may be caused by the following factors:

1. Failure of the company estimate to properly reflect the scope of work
2. Mistakes in the estimate
3. Estimator not taking into account current adverse market conditions
4. Unanticipated concern of bidders over availability of labor and/or materials, difficulty of terrain or site conditions, or the imposition of a stringent schedule in the bid package

If the cause can be corrected by clarifying the scope of work, extending the schedule, or approving alternate sources of labor, or other action then the bidders may be allowed to resubmit that portion of their proposals provided all of them are given the same opportunity to do so. If the situation cannot be corrected and the bidders have bid on the defined work, then proceeding or not proceeding will be a commercial decision and subject to a recommendation from the evaluation team.

UNIT RATES FOR WORK VARIATIONS

Once the bidders have, in competition, submitted their prices for a lump sum contract which includes unit rates for possible extra work, it is not customary for the client to attempt to negotiate a lower price or lower unit rates during the bid evaluation and clarification meetings. The same principle applies for day rates and fixed fees in cost plus contracts. However, all other portions of the bids are subject to clarification and review, particularly reimbursable cost items such as salaries and equipment rates.

Clients will attach considerable importance to unit rates offered in a lump sum quotation and intended to be used for variations to the work without the necessity to negotiate prices. Labor and equipment rates quoted in this manner will constitute all-inclusive payments, usually per hour, for extra work and will include all costs such as mobilization/demobilization, site overheads, camp and catering/maintenance costs, all direct labor costs, vacation pay, holiday pay, rest day pay, insurance, sick pay, completion and retirement bonuses, transportation, travel time to and from work sites, subsistence, living allowances, safety costs, small tools, consumables, and all burdens and indirect costs as well as profit, repairs, refuelling and lubrication, spare parts, operating supplies, depreciation, and taxes. Also, in certain cases, equipment will be inclusive of drivers or operators.

In addition to labor and equipment rates, clients may call for a quotation for "all-in" rates, which are unit prices for the provision of crews and equipment to carry out complete jobs such as electrical installation, trenching, or pipe laying per lineal foot or meter. Considering that it is normal for a large mechanical construction contract to attract 10% or 15% extra work through variation orders, great care is exercised by the client in the evaluation of this part of the lump sum bid.

The contract engineer is well advised to put as many variation unit price requests in the bid package as he can think of. This will help to avoid costly price negotiations later on during the course of the contract work. The bidder, on the other hand, should not volunteer any separate unit prices for extra work if the bid package does not ask for them.

Finally, in evaluation, the client may benefit from the advice of John Ruskin (1819–1900), the nineteenth-century essayist, critic, and reformer:

> "It is unwise to pay too much but it is unwise to pay too little. When you pay too much, you lose a little money—that is all. When you pay too little, you sometimes lose everything because the thing you bought was incapable of doing the thing you bought it to do.
>
> The common law of business balance prohibits paying a little and getting a lot. It can't be done!
>
> If you deal with the lowest bidder, its as well to add something for the risk you run and if you do that, you will have enough to pay for something better."

CHAPTER FIVE

Procedures and Manuals

CLIENT'S PROCEDURES

Project management, quality assurance, engineering, safety, contracts, procurement, construction, cost control, schedule control, personnel, administration, legal, insurance, accounting, and each and every discipline of the client's organization will be governed by the Project Procedures which should be written and produced at the very beginning of the project. This very seldom happens because all the key people are not on board at the beginning, particularly in construction, and those who are present are usually too busy to devote any time to this tedious but essential task. It is interesting to reflect that on most large projects there seems to be insufficient staffing in the early stages when huge numbers of contracts, manuals, and procedures are in simultaneous preparation, but at the tail end of the projects there is a tendency to hang on to people who haven't enough to do.

It may be thought that once a client has produced procedures for one project the same words will suffice for subsequent projects. In some disciplines, particularly administration and finance, this may be true, but for engineering and construction it is usually a completely new exercise. It must also be recognized that even the largest companies do not create superprojects at frequent intervals and when one comes along, any existing procedures from past projects are of limited use except as a guide for format.

It has been said, not without a certain amount of truth, that project procedures are a guide to the wise man and instruction to the fool. They certainly create employment, not only for the number of people who write, check, and approve them, but also for the army of QA/QC and contract compliance personnel who audit and enforce them. And yet, probably better than half of the total workforce employed throughout the life of the project never see them!

Procedures are descriptions of practices to be adhered to and definitions of associated responsibilities. They are necessary and valuable in the conduct of the work but publication should not be allowed to get out of hand to the extent that there is a procedure for every natural function. One well-known state-controlled oil company produced five massive volumes of procedures including a procedure on business ethics, conflict of interest, and security of information running to several pages, complete with charts. A single-page letter circulated to the staff would have done just as well.

A procedure should be produced to give direction on a subject or series of actions and to provide a model for the conduct of the work for each discipline. A procedure does not necessarily guarantee that the method laid down is positively the best method but it ensures that everyone is marching to the same drumbeat. There are, for example, many excellent and efficient ways of documenting variations in a contract scope of work but in a project where twenty or more contract engineers may be employed, it would be ridiculous to allow them to choose their own method of processing change orders, however experienced the individual may be. It is eminently preferable to have a standard Procedure A which everyone will follow but many consider to be inferior to Procedure B, than to allow the proponents to follow individual and alternative courses.

A procedure, therefore, should define the practices to be adhered to, demarcate responsibilities, describe the method to be employed, and provide references and exhibits if relevant. Procedures should not be set in concrete but should be revised from time to time if necessary to keep pace with the changing demands and conditions of the project. Unfortunately, some project managers hand down their procedures like tablets from the mountain and refuse to consider revisions even when it is plain that the original message is outdated.

Contract Administration Procedures

As mentioned, only a few project procedures are candidates for standardization; most of them are too closely aligned to the scope of work to conform with a standard. Contract administration procedures, however, lend themselves to this treatment and the following procedures could be standardized and presented for general use in the industry.

1. Contract administration
2. Work orders, change orders, and contract amendments
3. Claims

Here are some sample procedures for such activities. They would be part of the Project Procedure Manual along with the engineering, construction, finance, and other procedures.

Procedure No. 123

File No. 1.2.A.3.1
Date: June 5, 1993
Title: Contract Administration (Field)   Sheet No. 1 of 3

CRUDE OIL COMPANY INCORPORATED

Approved
Project Manager

O. Y. Knott

1. Purpose
2. Responsibilities
3. Administration
4. Backcharges
5. Meetings
6. Contract Close Out

1.   Purpose
1.1   The purpose of this procedure is to establish the manner in which the Project Team will administer all contracts following award.
2.   Responsibilities
2.1   The Contracts Engineer assigned to construction contract(s) is responsible for monitoring the overall performance of Contractor(s) in accordance with the terms and conditions of each contract. These responsibilities will include but will not be limited to:
2.1.1   Monitoring and expediting Contractor's performance and progress against the Master Schedule.
2.1.2   Ensuring that Contractor conforms with timely submittal of all procedures and documentation as required by the Contract.
2.1.3   Reviewing Contractor's reports.

2.1.4   Reviewing Contractor's pro forma invoices and submitting for payment.

2.1.5   Preparing and processing authorized Work Orders, Change Orders, and Amendments as required.

2.1.6   Conducting all job site meetings with Contractor.

3.   *Administration*

The Contracts Engineer will carry out the following administrative duties throughout the duration of each contract:

3.1   Maintain Contracts Department filing system in accordance with Procedure No. * * *

3.2   Review all contract correspondence and prepare correspondence to Contractor for Company Representative's approval and signature.

3.3   Prepare minutes of all job site meetings held with Contractor and distribute as required.

3.4   Transmit specifications and drawings and revisions to Contractor, as and when they become available.

3.5   Initiate and maintain a daily log of each Contractor claim at the earliest manifestation and commence a separate file on the claim complete with full backup information including but not necessarily limited to, records of the following data:

Progress, manpower, weather and equipment reports, schedules, including all revisions, drawings and specifications, including all revisions, minutes of meetings, including job explanation and bid clarification meetings, telephone conversations, relevant correspondence, memos to file, and photographs.

3.6   Prepare and process all Work Orders, Change Orders, and Amendments.

3.7   Prepare and distribute the following:

Monthly cash flow forecasts

Contract Status reports

4.   *Backcharges*

Backcharges may be raised on Contractor or Vendor when expenditure incurred by Company is to be recovered. Backcharges to Vendor of material or equipment may arise through incomplete or damaged material or equipment and although these backcharges are processed by the Procurement section, there will be close liaison with the Contracts Engineer through the effect on Contractor's performance caused by Vendor default. Backcharges to Contractor may be raised following equipment loan or provision of services by Company or in situations where work has been carried out by Company or by another contractor appointed by Company to remedy Contractor neglect such as failure to clean up the site or any other departure from the Scope of Work.

The Contracts Engineer will be responsible for the following actions:

4.1   Initiate and maintain a backcharge register.

4.2   Advise Contractor of the impending backcharge and raise the relevant debit.

4.3   Distribute copies of the backcharge debit to the Finance Department.

5.   *Meetings with Contractor*

5.1   The Contracts Engineer will preside over progress and negotiating meetings with Contractor and will prepare and distribute minutes of such meetings.

5.2   Meetings with Contractor will take place on Company premises or neutral premises.

5.3   In addition to progress meetings, the Contract Engineer will order regular meetings with Contractor to discuss and settle variation orders and to determine the existence of potential claims.

6.   *Contract Close Out*
       On completion of all contractual obligations by Contractor, the Contracts Engineer will carry out the following actions toward Contract Close Out:

6.1   Ensure that all necessary completion certificates have been processed.

6.2   Review and pass Contractor's final invoice to Finance.

6.3   Ensure that all backcharge debits have been settled.

6.4   Complete Contract Close Out checklist.

6.5   Prepare "Certification and Release" document.

6.6   Complete Contractor Performance Report.

Procedure No. 124

File No. 1.2.A.3.2   Sheet No. 1 of 2
Date: June 5, 1993
Title: Work Orders, Change Orders, and Contract Amendments.

CRUDE OIL COMPANY INCORPORATED

Approved
Project Manager

O. Y. Knott

1. Purpose
2. Responsibilities
3. Application–Work Orders
4. Application–Change Orders
5. Application–Contract Amendments
6. Exhibits (Note: see Appendixes for models)

1.   Purpose

1.1   The purpose of this procedure is to establish the manner in which the Project Team will modify contracts through the use of Work Orders, Change Orders, and Amendments.

2.   Responsibilities

2.1   The Contracts Engineer assigned to construction contracts is responsible for the application, preparation, documentation of supporting material, drawings, and the like, receiving, where relevant, Contractor's quotation for the extra work, obtaining Company "fair price" estimate, negotiation and recommendation for approval by Company Representative, processing, and distribution of Work Orders, Change Orders, and Amendments.

2.2   The Cost Engineer, when requested by the Contracts Engineer, will prepare a "fair price" estimate for the proposed extra work.

3.   *Application–Work Orders*

3.1   The Work Order, a model of which is presented in Exhibit 1. (Appendix 4) to this Procedure, will be used in situations where extra work is required to be performed within the general Scope of Work and unit or other prices for such work are clearly written into the Contract.

3.2   When signed by both parties, the Work Order formally requires Contractor to perform the modification in accordance with the terms and conditions of the Contract and at the prices contained therein. Work Orders will be issued at Lump Sum prices by application of the Contract rates for extra work against the agreed period and labor force required for the Work. Work Orders will not be issued with a "not to exceed' figure or in circumstances where Company does not wish to use the extra work rates in the Contract.

4.   *Application–Change Orders*

4.1   The Change Order, a model of which is presented in Exhibit 2. (Appendix 3) to this Procedure, will be used in situations where extra work is required to be performed within the general Scope of Work but clearly applicable unit or other prices are not available in the Contract or such prices are available but Company does not wish to use them.

4.2   Should the extra work proposed not be within the general scope of work, neither the Change Order nor the Work Order will be used but the work or modification will be implemented by means of a Contract Amendment.

4.3   Change Orders will have space for the signatures of both parties and will contain the following information:

4.3.1 Type of payment (fixed, estimated or "not to exceed").

4.3.2 Backup information and correspondence leading to the change.

4.3.3 Who initiated the change proposal? (Company or Contractor).

4.3.4 Scope of the Extra Work complete with drawings.

4.3.5 Reasons for the change.

4.3.6 Cost and Cost Codes.

4.3.7 Start and completion dates for the extra work.

4.3.8 Effect on baseline schedule, if any.

4.3.9 Whether Change Order represents an increase in the Contract Lump Sum or a decrease (negative change).

5.   *Application–Contract Amendments*

5.1   The Contract Amendment, a model of which is presented in Exhibit 3. (Appendix 2) to this Procedure, will be used as follows:

5.1.1 Where the proposed change does not fall within the general Scope of Work.

5.1.2 To change or modify the terms of the Articles of Agreement.

5.1.3 To alter terms in parts of the Contract other than the Scope of Work.

5.1.4 To alter the Contract Price or Unit Rates.

5.1.5 To settle claims.

5.1.6 To add or delete substantial sums related to the Scope of Work but which require more formal language and documentation not found in Change Orders.

5.1.7 To change the baseline master schedule, milestone dates, or completion date(s).

5.1.8   To add to the Contract Price when the total cumulative value of Work Orders and Change Orders reaches XX% of the value of the Contract as amended.

Procedure No. 125

File No. 1.2.A.3.3
Date: June 5, 1993
Title: Claims   Sheet No. 1 of 3

CRUDE OIL COMPANY INCORPORATED

Approved
Project Manager

O. Y. Knott

1. Purpose
2. Definition
3. Responsibilities
4. Handling and settlement of claims
5. Appointment of "Expert"
6. Arbitration
7. Legal proceedings

1.   Purpose
1.1   The purpose of this procedure is to establish the modus operandi in the receipt, examination, processing, and settlement of contractor performance claims against Company and the preparation and pursuance of counter claims against Contractor. This procedure is concerned only with Contract related claims. Claims for loss or damage to property, personal injury, or death are properly dealt with by Company's Insurance Department and do not form part of this procedure.
2.   Definition
2.1   A claim is a demand in writing from Contractor to Company for payment for services performed, expenditures incurred, or losses suffered which cannot be resolved within the terms and conditions of the Contract. A claim may arise from a variation order proposal initiated by Contractor for work which Contractor believes to be extra to the Contract Scope of Work but which Company declares to be part of that scope.
2.2   A counter claim is a demand in writing from Company to Contractor to deduct from the Contract Price or offset against a claim any monies which may be due to Company through services provided by Company which Contractor had agreed to provide within the Contract Price but had failed to do so.
3.   Responsibilities
3.1   The Contracts Engineer is responsible for receiving, acknowledging, and coordinating the processing of contract related claims.

3.2   The Company Claims Review Board is responsible for convening to attempt to reach settlement of claims when requested to do so by the Contracts Engineer. The Board shall seek the advice of the Law Department if appropriate.

4.   Handling and Settlement of Claims

4.1   The Contracts Engineer will be aware of a pending claim usually by receiving a variation proposal from Contractor for a Work Order or a Change Order for alleged extra work. Should the Contracts Engineer consider that such proposal is not justified, he will disallow the whole or part and attempt to reach agreement with Contractor to withdraw or negotiate that part not in dispute. If the disputed portion still remains, the matter will be registered as a claim.

4.2   Immediately upon the recognition of the existence of a claim, the Contracts Engineer will proceed to log the details as described in Procedure No. 123– Contract Administration.

4.3   If the amount claimed is less than _____ and within the authority of Company Representative to effect settlement, the Contracts Engineer will first convene a meeting of Project Team members involved in the claim including the Cost Engineer. This group will arrive at a recommendation as to the amount to be offered to Contractor by way of settlement and will refer the matter to Company Representative. Following negotiating meetings with Contractor, Company Representative will authorize payment within the limits of his authority.

4.4   If the amount claimed is greater than _____ and Contractor persists in its claim, the matter will be referred to the Claims Review Board who will receive all details and backup information on the claim from the Contracts Engineer and will request an estimate from the appropriate estimating unit to assist in assessing and resolving the claim. The C.R.B. will convene negotiating meetings with Contractor as necessary and will attempt to arrive at a satisfactory settlement in accordance with Company policy. In the event of a satisfactory settlement being reached and the amount settled is less than _____ , the Contracts Engineer will prepare a Change Order as a vehicle for settlement. If the amount settled is greater than _____ , a Contract Amendment will be processed to change the Contract Price accordingly.

5.   Appointment of "Expert"

5.1   Should Company and Contractor be unable to reach an agreement on a satisfactory settlement of a claim, Contractor may be invited to agree to the appointment of an "expert" to adjudicate in the matter. This arrangement is offered as a less expensive and less time consuming alternative to arbitration.

5.2   The expert will be a person of considerable knowledge in the field of construction related to the matter in dispute. Both Company and Contractor must reach agreement on the appointment and must be satisfied as to the complete neutrality of the candidate. The decision of the expert will be final and binding on both parties.

6.   Arbitration

6.1   Should Company or Contractor form the opinion that the dispute is too complicated or too important to rest on the decision of a single expert, the matter may be referred to Arbitration.

6.2   The initial step will be the filing of a request for arbitration with the appropriate Arbitration Institute. This body will determine the number of arbitrators, which is usually three. A list of names of possible arbitrators will be sent to the parties by the Institute. The parties will examine the list and will have the opportunity to delete the names of persons against whom they may have objections. Once this is done or if the parties have no objections, the arbitral tribunal will be formed. A hearing will take place where the parties may present their respective cases. The decision of the arbitral tribunal will be final and binding on both parties.

7.   *Legal Proceedings*

7.1   Should Company or Contractor not agree with the appointment of an expert or an arbitral tribunal, or either party becomes in breach of contract by failing to comply with the decision of the expert or the arbitral tribunal, or for any other reason the dispute remains unresolved, the matter may be the subject of legal proceedings. In this event, the matter on behalf of Company will be handled entirely by Company's Legal Department. All Project Team participants involved in the claim shall provide the necessary support to the Legal Department, as requested.

The Contracts Engineer, in addition to placing all relevant files and backup information at the disposal of the Legal Department, will obtain statements from all project personnel who may have been involved with the claim, for future possible use by the Legal Department.

OTHER PROJECT PROCEDURES

The following model procedures are representative of definitions and guidelines issued by other departments prior to project construction.

Notwithstanding the traditional links between contracts and procurement departments, particularly in the home offices, the contracts engineer in field conditions works more closely with the cost engineers than with any other discipline. It is only fitting, therefore, to include cost control and estimating procedures alongside those of contract administration. In addition, because of contractor involvement, the gathering of "as built" information is also included in the model procedures.

Procedure No. 126

File No. 2.1.A.1.1.
Date: June 5, 1993
Title: As Built Documentation   Sheet No. 1 of 3

CRUDE OIL COMPANY INCORPORATED

Approved
Project Manager

O. Y. Knott

1. Purpose
2. Definitions

3. Responsibilities

---

1.   *Purpose*
1.1   The purpose of this procedure is to define the practices to be adhered to and the associated responsibilities, in effecting a change to Approved for Construction engineering documentation during fabrication, construction, installation, and commissioning.
1.2   The procedures also define the practices to be adopted in the approval distribution, recording, and filing of the revised documents which will detail an "as built" system.
2.   *Definitions*
2.1   D.E.C.: Design Engineering Contractor
2.2   Vendor: A manufacturer, distributor, or fabricator who supplies a material, equipment, machine, component, or tool for a specific purpose.
2.3   Contractor: Shall mean any person or company entered into a contract to perform work and/or to provide a service.
2.4   Site Representative: A supervisor appointed from the Project to act on its behalf in monitoring, review, and coordination of the Contractor.
2.5   Site: The place or places where construction activities are carried out.
2.6   Engineering document: A document developed by the project organization that upon completion of a successful review cycle and subsequent approval becomes a justification for engineering development and/or description of work to be performed.
2.7   A.F.C.: Approved for Construction.
2.8   Site Inquiry: A technical query raised at site by either the Contractor or Site Representative requiring clarification on or revision to an engineering document by the Project team.
3.   *Responsibilities*
3.1   The necessity for maintaining a record of any deviations from A.F.C. documents subsequent to D.E.C. completion up to and including plant startup is fundamental to the successful completion of the project. The incorporation of these changes onto the aforementioned documents will provide a complete package of "as builts" and hence suitable for certification and records.
3.2   In order to maintain a common approach and format throughout all site locations when recording and processing engineering queries, the requirements of the relevant procedure governing site inquiries are to be fulfilled by the Contractor. The inquiry form is to be completed by the Contractor and actioned according to procedure requirements.
3.3   Where such a form details a proposed change to engineering design or philosophy, review and approval by project engineering is a necessary requirement.
3.4   Project engineering in conjunction with the associated D.E.C. or other responsible parties will review and recommend approval or rejection.
3.5   During the construction and installation phases it will be the responsibility of the Contractor to bring to the attention of the Site Representative any requirements for clarification on and/or changes to the A.F.C. documentation produced by the D.E.C.
3.6   This is to be effected by completing the Site Inquiry form.
3.7   This query will then be actioned by the Site Representative.

3.8   For every query and subsequent approved change the Contractor will be responsible for maintaining a record of changes to design documentation.

3.9   Where the query involves a revision change, this will be marked up on the site working print copy by the Contractor.

3.10   Subsequent to completion of the Scope of Work according to the Contract, the Contractor will then be responsible, within a time period to be agreed with the Company, for a complete set of engineering documents marked up with all revisions and changes, including material changes and certifications and where documents have not been subject to change, this must be clearly marked on the front sheet or title block.

3.11   The D.E.C. will then be responsible for incorporation of these changes to the original documentation.

Procedure No. 127

File No. 2.1.A.1.2
Title: Cost Control   Sheet No. 1 of 6

CRUDE OIL COMPANY INCORPORATED

Approved
Project Manager

O. Y. Knott

1. Purpose
2. Definitions
3. Responsibilities
4. Descriptions

1.   Purpose

1.1   To outline the basic methods and requirements for Project Cost Control and to explain the types of Contract Values, Performance Analyses, and Cost Forecasts used.

2.   Definitions

2.1   List of Abbreviations

ACWP   Actual cost of work performed
AFE   Approval for expenditure
BCWP   Budgeted cost of work performed
BCWS   Budgeted cost of work scheduled
CCE   Current control estimates
CCV   Current control value
CO   Change order
CPR   Cost performance ratio
CWE   Current working estimate
CWO   Contract work order
ECV   Estimated contract value

ITC   Indicated total cost
MCE   Master control estimate
OCV   Original contract value
PCE   Preliminary control estimate
PCF   Project control forecast
SPR   Schedule performance ratio
WBS   Work breakdown structure
WP   Work package

2.2   Work breakdown structure (WBS). A subdivision of the complete project work scope which will:

(a)   Identify the major end items or products needed to accomplish the project objectives.

(b)   Provide the necessary framework for the establishment of Project Baselines for cost, time, and resource.

(c)   Provide a structure for orderly summarization of work performance to selected levels of detail.

(d)   Assign responsibility for task accomplishment to specific organizations or persons.

The WBS is subject to revision every six months to reflect the current situation and revised contract strategy.

2.3   Work Package (WP). A defined work element or task which is clearly identified in the WBS as a discrete level of work where cost, schedule, and/or resources can be expressed.

2.4   Project baseline. Includes an identification and organization of the work scope and a cost estimate and schedule agreed at completion of the pre-engineering phase. Normally, the following documents would be included:

(a)   Work breakdown structure
(b)   Work breakdown structure description sheets
(c)   Master control estimate
(d)   Master control schedule

2.5   Project calendar. The number of calendar weeks during the defined project duration.

2.6   Measurable work. Items or areas of the work scope that are quantified and progress related.

2.7   Non Measurable work. Items or areas of the work scope that are indirect and/or not progress related.

2.8   Contractors baseline. This is similar to Project baseline but applies to a particular Contract and Contractor. It includes an identification and organization of the work scope and a cost estimate and schedule agreed at Contract award. Documents involved include:

(a)   A work breakdown structure
(b)   Estimated Contract value
(c)   Contractor's schedule

2.9   Project control estimates.

2.9.1   Preliminary control estimates (PCE). These are based on the preliminary project baselines and prepared from conceptual data before the pre-engineering design is complete.

2.9.2   Master control estimate (MCE). This is the basic reference cost estimate produced at the end of the pre-engineering phase. It is built up in the WBS format by calculating costs for each individual work package.

MCE = Base estimates + Escalation + Contingency

2.9.3   Current control estimate (CCE). Based on revisions to the WBS, this estimate is approved by management based on:

(a)   Current information from the various technical disciplines
(b)   Base estimates for future contracts/purchase orders
(c)   Bid evaluations
(d)   Contract awards
(e)   Forecasts to complete

The CCE replaces the MCE for cost control and reporting purposes and is revised as required but normally every six months.

2.9.4   Check estimates. These are prepared prior to Contract award and based on the relevant bid package. They are used for bid evaluation and control of CCE accuracy.

3.   *Responsibilities*

3.1   The project controls group is responsible for cost control generally including cost reporting against approved control estimates and Contractors' baselines, logging and analyzing changes, compiling cost forecasts based on current information and cost appraisals, and maintaining the work breakdown structure and cost code systems of organization.

3.2   Site surveillance is responsible for cost control at the work location including monitoring and analyzing costs as work proceeds, reporting to project cost controls group, and instituting and monitoring corrective action.

3.3   Project management is responsible for approval of contract values, commitments, etc., review of potential changes and cost forecasts, and direction of corrective action where appropriate.

4.   *Description*

4.1   Basic objectives

4.1.1   Cost control is concerned with anticipating future costs in conjunction with the tracking of ACWP. Accurate monitoring of costs, as work proceeds, provides a basis for forecasting final costs. By definition, ACWP is cost already incurred and cannot be altered whereas forecast costs can be influenced by management action. In this respect cost control helps maintain budgeted costs and gives early warning of deviations from the established plan, enabling corrective action to be taken where appropriate.

4.2   Contract values

4.2.1   Contract values represent the established and approved value of a Contract at any given time and are involved in the formulation of project control estimates, cost forecasts, AFE, and commitment values.

4.2.2.   The original contract value (OCV) is the initial value of a Contract, agreed between the Contractor and Company, based on the defined Contract Scope of Work at the time of Contract award.

4.2.3   The estimated Contract value (ECV) adds allowances to the OCV for provisional sums, to be used as required and a calculation of the effects of escalation over the Contract period.

4.2.4   The above Contract values are obtained by way of a commitment value report prepared by the contracts department, at Letter of Intent stage and agreed with project cost controls.

4.2.5   As a contract progresses, the current contract value (CCV) is developed based on the ECV to reflect approved changes. The CCV is prepared to project cost control group and is used to adjust commitment values.

4.2.6   The OCV and ECV are fixed values set at the start of a Contract while the CCV will change as the Contract progresses.

4.2.7   The above values apply to committed Contracts only. For noncommitted work packages the approved control estimate forms the base value until such time as a Contract is awarded.

4.3   Potential changes

4.3.1   Although a potential change will require careful consideration and evaluation in any new forecast, it does not in itself constitute a new forecast nor does it substitute for the change control process as stipulated in Procedure 456.

4.3.2   The project cost controls group will review and analyze potential changes and make suitable allowances for forecasting purposes. Those of a significant nature will be brought to the attention of project management.

4.3.3   Corrective action, if necessary, will be determined by project cost controls group or project management, depending on the magnitude of the potential change concerned, after due analysis and discussion.

4.3.4   Corrective action may be any modification that will bring the potential change into line with scope, budget, and schedule, or results in a minimum adjustment to these. If a potential change has a significant impact after corrective action or if no suitable corrective action is available, it must be accepted by estimating the total consequences and including this in new forecasts or, if there is a resultant change in Contract scope, by initiating a Change Order in accordance with project procedures.

4.4   Cost appraisal

4.4.1   A cost appraisal involves trends and performance analyses.

4.4.2   Site surveillance will monitor trends against the established baselines and cost plans and carry out the following performance analyses:

(a)   Budgeted cost of work performed (BCWP)
(b)   Budgeted cost of work scheduled (BCWS)
(c)   Actual cost of work performed (ACWP)

4.4.3   Project cost controls group will review the above and carry out the following performance analyses:

(a)   Cost performance ratio (CPR)
(b)   Schedule performance ratio (SPR)
(c)   Cost variance
(d)   Manhour cost

4.4.4   Information gained by the cost appraisal will be used in the development of cost forecasts.

4.5   Cost forecasting

4.5.1   The purpose of forecasting is to provide an estimated final cost based on information and experience accumulated, up to the forecast date, by the participating groups within the project.

4.5.2   Forecasting is carried out at Contract level by project cost controls group using progress information and analyses provided by site surveillance. The CCV forms the basis for calculation of final costs and there are three distinct phases involved in the process.

4.5.2.1 The indicated total cost (ITC) is the result of the first phase formed by the addition to the CCV of the most probable cost items, i.e., pending changes and potential changes.

4.5.2.2 The project control forecast (PCF) forms the next phase by adding to the ITC allowances based on the Contract cost appraisal and any claims situations which may have arisen, thus representing the most probable final cost.

4.5.2.3 The current working estimate (CWE) is the third phase and represents the maximum foreseeable costs at any given time based on the PCF with allowances for future potential changes.

4.5.3 The above forecasts are reported to project management for action where appropriate. Such corrective action as designated by project management will be implemented by site surveillance and subsequent progress will be monitored and reported to project cost controls group.

4.5.4 The CWE is used in CCE and AFE updates and contingency release where appropriate.

4.6 Monitoring and reporting

4.6.1 To achieve effective cost control Contracts must be closely monitored and information analyzed to produce control data.

4.6.2 It is the Contractor's responsibility to develop baselines and cost plans and to report progress in an accurate and timely fashion, providing cost performance and cost management reports.

4.6.3 Site surveillance will review, check, and analyze Contractor provided information and report to project cost controls group.

4.6.4 Reporting will include the following:

(a) An assessment of progress against the established baselines and cost plans

(b) Status of change orders

(c) Cost appraisal

(d) Advice of potential changes with classification

(e) Status of corrective action currently in progress

(f) An assessment of claims where appropriate

4.6.5 Project cost controls group will utilize such information to establish current project status and develop cost forecasts. Status reports and forecasts will be forwarded to project management for review and action where necessary.

Procedure No. 128

File No. 2.1.A.1.3
Date: June 5, 1993
Title: Estimating   Sheet No. 1 of 3

CRUDE OIL COMPANY INCORPORATED

Approved
Project Manager

O. Y. Knott

2

1. Purpose
2. Definitions
3. Responsibilities
4. Description

1.   Purpose
1.1   To outline the basic objectives of project control estimating and to describe the stages in the estimating process and functions within the overall control system.
2.   Definitions
2.1   For all definitions and list of abbreviations refer to Procedure No. 127– Cost Control.
3.   Responsibilities
3.1   Project cost controls will prepare, maintain, and revise all project cost control estimates and AFEs as required and report to project management.
3.2   Project management is responsible for approval of all project control estimates and AFEs and designation of corrective action where necessary.
4.   Description
4.1   Basic objectives
4.1.1   Estimating is a means of predicting the project costs, and forms the cost measurement baselines for comparison purposes. Cost monitoring and reporting reflect the actual costs to date. Forecasts based on actuals are compared with the current estimate, and corrective action to be taken is suggested where appropriate.
4.1.2   To initiate an estimating exercise a "Request for estimate" is prepared by the estimating group and approved by the Cost/Estimating supervisor.
4.1.3   All estimates will be kept in a secure place and files maintained by the estimating group.
4.2   Preliminary control estimate (PCE)
4.2.1   The PCE is prepared from conceptual data in the early stages of the project to give an overall base value and enable initial budget values to be established.
4.2.2   At the time of compilation of WPD sheets, each work package is allocated an estimated cost using the information available. Base estimates are used in the formulation of project control estimates by raising to Contract level and subsequently to project level.
4.3   Master control estimate (MCE)
4.3.1   The MCE represents the basic project investment cost estimate and takes into account the following:
(a)   Project schedule
(b)   Pre-engineering and design drawings and information
(c)   Scope as defined in the WPD sheets
(d)   Preliminary control estimate
(e)   Tender documents and Contracts existing at the time of preparation.
4.4   Check estimates
4.4.1   A bid evaluation provides a firm basis against which to check incoming bids. Check estimates are compiled as required, using the same information as that supplied to the Contractor or supplier. This is normally carried out before bid packages are released or during the Contractor's or supplier's bid preparation.

4.5　　Current control estimate (CCE)

4.5.1　Where the MCE represents a fixed base cost reference, the CCE is a dynamic cost reference, developed from WBS revisions, Contract awards, Contract changes, and project forecasts.

4.5.2　The CCE takes into account the following current information at the time of compilation:

(a)　　Bid evaluations
(b)　　Check estimates
(c)　　Contract award data
(d)　　Baseline revisions

4.5.3　CCE updates are based on the CWE which is regularly revised to reflect current status. The CCE is revised as necessary but normally every six months.

4.5.4　Project cost controls group prepares and distributes all CCE updates as required. Each estimate, or revision to same, is put to project management for approval.

4.5.5　Where a revised estimate exceeds the AFE value, it must receive the approval of Company.

4.6　　Approval for expenditure (AFE)

4.6.1　AFE values are developed at Contract level, generally from the approved MCE or CCE and issued prior to Contract award. Each AFE contains a portion of the total escalation and contingency allowances. Subsequent revisions to the CCE are used as a basis for revising AFEs.

4.6.2　AFE values are prepared by project cost controls group.

4.7　　Baseline revisions

4.7.1　Adjustments to the existing baselines are produced at Contract level by the appointed Contractor and agreed with the appropriate site surveillance team.

4.7.2　Project cost controls group use the adjusted baselines for baseline revisions, with the addition of approved change orders and escalation.

CONTRACTOR'S PROCEDURES

The Work Procedure

The terms of the construction contract require the contractor to produce a work procedure for each phase of construction before that phase is scheduled to commence. The contractor's work procedures (variously job instructions, work instructions, job procedures, etc., depending on the nomenclature in use at the time and place) are, together with the contractor's Quality Assurance manual, his detailed description of how the work is to be performed.

Work other than administration and mobilization should not be allowed to commence until the work procedures are presented and approved by the client. On his part, the client should make it absolutely clear at the bid stage exactly what is required in the procedures as the severity of these requirements vary considerably depending on the client, the country, and the nature of the work. For the same reasons, the cost of work procedure production can also vary from a moderate sum to a very large amount involving special staff, word processors, consultants, and a major proportion of time and effort. Traditionally, all bidders on a large project tender are required to produce samples of the proposed QA manual and work procedures gratis. On critical contracts it may be worthwhile to require submission of such documents with the bids (or at the least a substantial outline) with an offer to pay the unsuccessful bidder a fixed sum for this work. Indeed, contractor work procedures for highly specialized undertakings, such as deep water pipe laying from sophisticated lay barges, are often priced out as part of the response to the invitation to bid. The unsuccessful bidder will receive a previously agreed sum for his efforts in producing the work procedure, quality assurance, and quality control manuals. The successful bidder, of course, will include the cost of these publications in the contract compensation section. Unfortunately for the contractor engaged in other construction ventures, the amount of work required in the production of such manuals is never fully explained at pre-award conferences, or given much importance until the kickoff meeting. Contractors have been known to collapse in a state of shock when they are given the eventual explanation of what is expected by way of quality and quantity of the work procedures, especially when work is not allowed to start until the procedures have gone back and forth between contractor and client many times before final approval.

The following sample work procedure describes the excavation, backfill, and final landscaping of a trench. For the purpose of this exercise, it is unimportant to dwell on the objects to be interred in the trench since the model is intended to illustrate what some clients may expect from such a procedure.

Crude Oil Company Terminal Project

Section 1. Civil Works
Revision X
Date of issue: 06.05.93   Page 1 of 3
DITCH DIGGERS INTERNATIONAL INC.

Work Procedure

Table of contents

1.12.1   Locations
1.12.2   Contours
1.12.3   Dump areas
1.12.4   Fences
1.12.5   Top soil
1.12.6   Revegetation
1.13   *Inspection of civil works*
1.13.1   ROW and grade
1.13.2   Trenching
1.13.3   Backfill
1.13.4   Clean up and landscaping

Specifications

SP-AA-123 Rev. A	:	General specifications
SP-AB-124	:	Concrete construction

Drawings

D-001-01	:	Road crossings
D-001-02	:	Drainage channels
D-001-03	:	Existing pipe crossings
D-001-04	:	Existing cable crossings
D-001-05	:	ROW restoration & drainage
D-001-06	:	Typical water stops
D-001-07	:	Typical ditch dimensions
D-001-08	:	Typical main road crossings
D-001-09	:	Typical bog area crossings
D-001-10	:	Typical ROW dimensions
D-001-11	:	Typical overhead telephone line and high-tension line crossings
D-001-12	:	Ditch route map from East point to West point (1:1000)

Standards

STD-1234	:	Concrete structures-construction and control
STD-5678	:	Methods for testing concrete
STD-9101	:	Safety in trench excavations

Data sheets

DS-011-01	:	Coordinates for traverse stations and bench marks
DS-011-02	:	Coordinates and horizontal partial distances of turning points

Crude Oil Company Terminal Project

Section 1. Civil Works
Revision X
Date of issue: 06.05.93   Page 1 of 13
DITCH DIGGERS INTERNATIONAL INC.

Work Procedure

1.1   *Setting out*
1.1.1   Checking centerline
        The data sheets DS-011-01 and DS-011-02 will be used for the setting
out and checking of anglepoint, angles, and levels of the centerline.
        The equipment for the setting out will be normal survey equipment;
Theodolites and Distomat.
        Reference points will be marked outside work area and identified by
coordinates.
1.1.2   Longitudinal profile
        The longitudinal profile is to be taken over the centerline.
        Profile will be drawn up in scale 1:200 on Format A3.
        The points where cross sections are taken are to be indicated on the
profile using wooden stakes with the centerline in the form of a cross. Stakes will
be 3 ft. in length at a distance from one cross section to another, depending on
variations in the longitudinal profile, not more than 200 yards.
        Cross sections are marked outside the work area with temporary markers
to retrieve measurement.
1.1.3   Cross sections
        Following determination of the station of the cross sections, the cross
sections will be drawn up in a scale 1:100 on Format A3.
        Cross sections are identified by number plus distance from one towards
West point.
        Both longitudinal profiles and cross sections will be verified by the use
of a stamp as illustrated below:

	FOR COCTP	FOR DDII
CROSS SECTION		
BOTTOM TRENCH		
TOP CONSTR. ROAD		
ROCK LINE		

(Step to be used for identification)

The longitudinal profile and the cross sections are for measurement purposes only.

1.1.4   Bottom of trench and top of construction road

The senior engineer is responsible for putting the bottom of trench line on the longitudinal profile. Trench and construction road lines will also be shown on the longitudinal profile. Both lines will be signed for on the stamp reproduced above.

These lines are for measurement purposes only.

1.1.5   Determination of rock line

If, in certain areas, rock is covered by silt and/or peat, this will be removed where trench and construction roads are to be established.

Line shall be signed for measurement purposes on the stamp shown above.

1.1.6   Information for drilling and blasting

On every cross section a white plate 12″ × 6″ and attached to a stake will be placed, bearing the following information:

1. Level indicating top of plate
2. Depth of trench in inches
3. Distance in feet between center line of trench and roadside at trench

The marker will be placed in the centerline of trench.

1.2   *Right of way principle*

1.2.1   Preparation of ROW

In principle, the ROW is a strip of land maximum 100 feet wide and situated so that 60 feet is at the side where the construction road is to be and 40 feet at the other side of the centerline.

The working area is a strip 50 feet wide, particularly including the construction road but also the trench.

For each section and prior to commencement of work, a plan will be prepared. The plan will show proposal encroachments and resulting landscape requirements and all activities will be within limits for terrain encroachments shown on work area plans.

1.2.2   Soils handling

The ROW crew will carry out brush clearing, removing the organic topsoil and storing it along the working area. In rock area, the subsoil will then be removed from the trench in order to render it suitable for drilling. Topsoil and subsoil will be stored separately. Soil will be stored according to the working plan for the actual section.

1.2.3   Determination of rock line

The depth of the topsoil will be established by using a thin reinforcing rod or, alternatively, with a backhoe.

1.3   *Water disposal*

1.3.1   Method

In wet areas or after rainfall, the water will be disposed of in the following manner, using the alternative methods as applicable:

(a)   Surface water will be drained away at the side of the construction road. If necessary, a ditch will be excavated.

(b)   Water in trench will be pumped away into the above mentioned ditch or into existing waterways.

(c)   Where existing waterway is crossing the trench, rigid steel pipes will be used to carry water across at the top of the trench.

1.4   *Drilling and blasting*
1.4.1   Mode of operation
        The drilling and blasting crew will follow the ROW crew. Drilling will be performed to a loading and drilling plan.
        The blasting pattern and loads will be adjusted as necessary to stay within specification and obtain a degree of fragmentation that makes the rock suitable for filling in the construction road or backfilling the trench.
        The type and amount of explosive will depend on local geology.
        For their measurement, the drilling and blasting crew uses the information prepared by the setting out crew as given on the plates mentioned in the section on "Setting out," subsection 1.1.6.
        Before blasting in areas where other constructions may be damaged or for general safety, the rock surface will be covered by net or fiber cloth in order to prevent falling stones. General fragments falling outside the limits will be removed as a party of the landscaping.
1.4.2   Safety measures
(a)   Before blasting
        Five minutes of siren sound.
        Guards will be posted as necessary depending on the surrounding area.
(b)   After blasting
        Three short signals with intervals between the signals.
1.5   *Construction road*
1.5.1   Construction road materials
        Construction road will be built of rockfill from blasting inside working area. In area where construction road must be built before drilling and blasting in trench, rockfill from main construction dump area will be used.
1.5.2   Construction road in bog areas
        In bog areas, fiber cloth will be used under the construction road. In minor bog areas, the peat will be replaced by rock. This will be done according to approved working plan.
1.5.3   Crossing of trench
        Where the construction road crosses the trench, bridges of prefabricated slabs will be used.
1.6   *Trenching*
1.6.1   Type of trench
        For the purpose of this work procedure, trenches will be classified as follows:
        Type (a) Trench in solid rock
        Type (b) Trench with rock bottom where the rock is covered by silt and/or peat
        Type (c) Trench completely in silt and/or peat
1.6.2   Type (a)
        The size of Type (a) is determined by blasting. Trenches in rocky ground will be carefully finished, eliminating all projections and roughness and at the same time removing surplus rock fragments from the trench bottom. Where the trench is overblasted, the rock fragments in trench bottom will be compacted with a mechanical road roller to produce a bottom free from projections.
1.6.3   Type (b)

The same instructions apply for Type (b) trenches after removal of top soil, silt, and/or peat.

1.6.4   Type (c)

Type (c) trenches may be excavated closer to other working parties.

1.6.5   Excavated material

Excavated material will be used for building the construction road or stored at the side of the trench to be used for backfill according to the civil construction plan given in advance for the different cross sections, and depending on the quality of the material. Surplus material will be transported to the dump area.

1.6.6   Quality control

The survey crew will check the level and width of the trench bottom. Special attention will be paid to assure that the correct cover and width is obtained in particular at:

(a)   Bends

(b)   Road crossings

(c)   Utility crossings

The QC Inspector will inspect the trench visually and should any adverse comments arise, the necessary steps required for rectification will be taken.

1.7   *Laying bed*

1.7.1   Start of work

Laying bed will be installed under the object to be buried. Work will not start before inspection and approval.

1.7.2   Specification

The laying bed will consist of a 4″ layer of gravel 1/4″–1″ and a 4″ layer of 0″–1/4″ sand. Both layers will be compacted by a mechanical road roller. The top of the layers will be marked out prior to the laying operation.

1.7.3   Quality control

The laying bed will be inspected by QC inspector.

1.8   *Backfill*

1.8.1   Methods

Backfill will be executed by backhoe.

Sagbends will be backfilled and compacted before overbends in straight sections.

Where compaction is required, the layers will not exceed 12″.

1.8.2   Quality control

The surveying crew will check the top of surrounding backfill and will make a longitudinal profile.

QC Inspector will check each layer.

1.8.3   Materials

The material will consist of well graded sand and gravel 0″–1/4″ at horizontal bends, road crossings, and stream banks around the buried object. Following this, a first layer of pebbles and rubble mixed with sand and loose earth will be used together with material from the trench. The final layer will consist of material not larger than 12″, free from peat and organic soil.

1.8.4   Road crossings

The backfilling will be compacted at horizontal bends, road crossings, and stream banks. The layers will not be more than 12″ deep. Small hand-operated compactors will be used.

Road crossings will be carried out in accordance with Drawing D-001-01 and the final layer will be covered with a layer of asphalt where previously asphalted.

Road crossings will be checked and approved by the civil inspectors.

1.9   *Land drains*

1.9.1   Specifications

The drainage system will be installed in accordance with Drawings: D-001-02 and D-001-05. Minimum fall of the ditch will be 1:100.

1.9.2   Drainage

Where drainage piping already exists, the trench will be excavated with care in order not to damage the existing work. Damaged pipe will be replaced before installation of buried object.

Concrete drain pipes without sealed joints will be installed. When depth of trench is greater than 15 feet, reinforced concrete pipes will be used.

Fine chips 1/2"–3/4" will be layed around the concrete pipe. Minimum cover for the sides and top will be 12".

1.9.3   Drainage of working area

Drainage of working area will be carried out in accordance with drawing D-001-02.

The trench will be blasted and excavated 3 feet wide in order to obtain sufficient cover sideways.

1.10   *Water stops*

1.10.1   Specification

Water stops will be located and constructed in accordance with Drawing D-001-06.

The water stops will consist of woven PVC bags filled with fine sand. The bags will be placed to create a convex with the curvature oriented upstream.

1.10.2   Installation

In rocky ground, loose material will be removed by excavator with the last small pieces being taken out by hand in order that sandbags may be laid directly on rock surface.

The first 9" of the water stops will be laid down at the same time as the laying bed.

Backfilling on both sides of the water stop will be compacted.

1.11   *Concrete construction*

1.11.1   Specification

Concrete construction will be carried out in accordance with:

Specification SP-AB-124

Standards STD 1234 and 5678

1.11.2   Application

Concrete will be placed in the formwork using concrete pump, crane, or directly from the truck mixer. Pouring will not commence until the formwork and dimensions are inspected by the senior civil engineer.

The concrete will be vibrated.

The fresh concrete surface will be treated as required by tamping and steel trowelling. Curing will be secured by use of water or by using a membrane when necessary.

1.11.3   Formwork

Formwork will be constructed in material specified on relevant drawings and/or in Bill of Quantity. Dimensions will be in accordance with required tolerance class.

Formwork will not be removed until the concrete has attained the required strength.

1.11.4 Reinforcement

Reinforcement will be in accordance with the relevant drawings and standards mentioned above.

Reinforcement bars will be placed as shown in the drawings. Reinforcement bars will be free of any coating which may reduce bond with the concrete. To ensure the specified cover of the reinforcement bars, plastic spacers will be used. Should reinforcements have become contaminated, they will be cleaned using the appropriate technique or dumped.

1.11.5 Positioning

The surveying crew will set out for cast-in parts in accordance with the drawings. After the parts are placed, it will check that they are within tolerances. The cast-in parts will be secured in their positions by nailing to the formwork or welding to foundation bolts.

1.11.6 Firm rock foundation

If the foundation is to firm rock, all loose material will be removed by excavation, hand digging, and finally cleaned by water/air jet.

1.12 *Clean up and landscaping*

1.12.1 Locations

Landscaping will be carried out at the work area, disposal and storage areas, and access roads.

Landscaping will return the landscape as close as possible to its original shape. All waste will be removed.

1.12.2 Contours

Generally, cutting edges will be rounded off and filled to rounded edges using subsoil and/or surplus rock stored along the ROW and in construction storage area.

Sharp edges or rock will be removed by blasting.

In areas where the construction road is filled higher than the original terrain, the edge will be rounded off.

Surplus spoil will be graded in the terrain where possible. In flat areas where this is not possible, the surplus spoil will be used in landscaping elsewhere or transported to dump areas.

1.12.3 Dump areas

Edges of dump areas will be rounded off in the same way as mentioned in 1.12.2 above or graded into the side terrain.

1.12.4 Fences

Fences destroyed by installation works will be restored.

1.12.5 Topsoil

Following completion of the above mentioned work, the organic topsoil stored along the ROW will be graded on the top.

1.12.6 Revegetation

Revegetation will be carried out Specification SP-AA-123 Rev. A.

1.13   *Inspection of civil works*

1.13.1   ROW and grade

Inspection shall be carried out to:

(a)   Determine that necessary permits have been obtained before commencement of clearing and grading activities

(b)   Ensure contacts with landowners and/or tenants before any fence is cut

(c)   Check that all openings made and other fencing details are in accordance with specifications and where practical, in accordance with the wishes of landowners and/or tenants

(d)   Ensure that the centerline, clearing limit, grain and other stakes are set at sufficiently close intervals to accurately locate and set out the required construction

(e)   Check that temporary offset stakes and reference bench marks are provided as required for rapid and accurate checking of locations and elevations

(f)   Check construction staking for sufficient accuracy to ensure installation with tolerances as specified or shown on drawings

(g)   Ensure that no false fill is deposited in the ditch line

(h)   Ensure that spoil, brush, and debris are disposed of in the correct manner

(i)   Ensure that preservation of ROW boundary markers and that ROW clearance is kept minimal

(j)   Report in detail any off-ROW damages to private property, roads, fences, and the like, and photograph such damage as appropriate

(k)   Ensure that all existing structures that interfere with construction are protected, relocated, or removed as required

(l)   Ensure that temporary drainage facilities are constructed in accordance with specifications where grading interferes with natural or previously installed drainage

(m)   Ensure the maintenance of required retaining structures against possible flooding and/or restriction of flow in natural channels

(n)   Determine that required permits are obtained before any blasting operations commence

(o)   Ensure that all applicable safety measures, codes, and restrictions are compiled with during blasting operations and that any debris is removed and disposed of in accordance with specified requirements

(p)   Prepare daily reports and submit these along with any other required documentation.

1.13.2   Trenching

Inspection shall be executed to:

(a)   Ensure the maintenance of up-to-date copies of alignment sheets, construction line lists, and permits on the ROW and approved work plans

(b)   Determine that all restrictions and permit conditions are complied with

(c)   Report in detail any off-ROW damages to private property, roads, fences, and the like, and photograph such damage where appropriate

(d)   Determine that centerline stakes and offset reference stakes and markers are provided at sufficiently close intervals as to permit accurate positioning of the ditch

(e)   Ensure that offset reference stakes and markers are preserved at all times during construction

(f)   Determine that existing underground structures and lines are located and protected

(g)   Ensure that excavation of the trench is centered on the stake centerline unless otherwise shown on construction drawings

(h)   Check trench dimensions against typical trench section dimensions shown on construction drawings

(i)   Ensure that overexcavation is corrected and ditch bottom is filled to the required level

(j)   Determine that placement, grading, and compaction of bedding material is in accordance with specified requirements

(k)   Ensure that the ditch walls are supported and sloped where necessary to prevent cave-ins and sloughing

(l)   Ensure that any ditch left open is in accordance with specifications, contract documents, and applicable government codes

(m)   Ensure that temporary bridges of adequate strength and dimensions are provided as required in locations where ditches cut access roads, paths, walkways, or similar facilities, and may interfere with normal traffic

(n)   Ensure that proper warning signs and markers are provided and maintained

(o)   Report all rock trenching and be familiar with applicable laws, codes, safety requirements, and techniques

(p)   Determine that required permits are obtained before any blasting operations commence

(q)   Ensure that all applicable safety measures, codes, and restrictions are complied with during blasting operations and that any debris is removed and disposed of in accordance with specified requirements

(r)   Monitor operations at road and other crossings to ensure that depth of cover as well as other requirements comply with specifications

(s)   Prepare daily reports and submit these together with any other required documentation.

1.13.3   Backfill
       Inspection shall be carried out to:

(a)   Ensure that backfill material contains no rocks larger than allowed by specifications

(b)   Measure the amount of cover both before and after backfilling, as necessary, to determine that the required depth of cover has been achieved using transits, probes, or other devices as required

(c)   Ensure that any required compaction of backfill material is performed in designated locations and check to ascertain that compaction meets the specified requirements

(d)   Ensure that agreements made with landowners and/or tenants for waste disposal are complied with

(f)   Prepare daily reports and submit these together with any other required documentation.

1.13.4   Clean up and landscaping
       Inspection shall be carried out to:

(a)   Be familiar with requirements for final grading, fencing, marking, backfilling, and revegetation of the ROW, and maintain copies of specifications and other controlling documents for reference during clean up and revegetation work.

(b)   Ensure that grading of the ROW is completed in accordance with specifications and that any required terraces and breakers are installed as specified

(c)   Ensure that in areas where the topsoil has been set aside it is used for the final backfill layer

(d)   Ensure that all foreign material, stumps, debris, and rocks larger than allowable size are removed from the backfill material

(e)   Prohibit the practice of burying debris on the ROW or in the backfill

(f)   Verify that all temporary works such as access roads, diversion ditches, and dams are removed and the sites restored

(g)   Ensure that all tiles and drains which have been broken or removed are replaced as required

(h)   Verify that all temporary gates are removed and fences replaced in accordance with specified requirements

(i)   Verify that ground and aerial markers are set as specified

(j)   Ensure that all vents, markers, and other appurtenances are painted as specified

(k)   Monitor sodding, seeding, and planting activities to verify that all revegetation is accomplished in accordance with specified requirements

(l)   Ensure that agreements made with landowners and/or tenants, as well as specification requirements for waste disposal, are complied with

(m)   Ensure that the necessary and required acceptance signatures from landowners and/or tenants are obtained

(n)   Prepare daily reports and submit these together with other required documentation.

Note: Sample Specifications, Drawings, standards, data sheets, and forms are not included in this model contractor work procedure.

MANUALS

The Quality Manual

The Project Quality Manual produced by the contractor will include the quality assurance and quality control programs. The format and content of this manual should cause no problem for the experienced contractor and the client will expect that a firm written policy on quality is already in existence and an established part of the contractor's organization. The contractor's quality assurance program should provide the client with confirmation that adequate measures have been taken to perform the assigned construction activities in accordance with the terms, conditions, and requirements of the contract documents. Quality control is the management of the activities of the contractor and its suppliers and subcontractors to comply with the requirements of the contract specifications.

The client will usually insist that the QA/QC Manager have a direct line of approach to the Contractor's Representative or whoever is the contractor's senior manager on the project. The QA/QC Manager will plan and maintain the quality programs and quality audits and will have the authority and responsibility for assuring that all phases of the Quality Manual are implemented. He or she will also be empowered to stop the work in areas which have a quality problem, in the same manner as the Safety Officer will halt unsafe practices until they are rectified.

The manual itself will usually begin with a declaration of the contractor's policy on QA/QC and will explain the various responsibilities of contractor's staff in connection with the implementation of the manual requirements. The manual may contain one or two pages of relevant definitions and also include the contractor's organization chart for its QA/QC department. The manual will deal in sufficient detail with document control and distribution, materials, and process control including certain aspects of purchasing, source inspection, material receipt and issue, and identification. The remainder of the manual will be concerned with the technical details of the work according to the contract and will conclude with descriptions of reporting and auditing methods, corrective actions, and the transfer of final records and "as built" documentation to the client.

The Project Proposal

There is one other volume issued by or on behalf of the client, perhaps by the design and engineering contractor. It is made available to the more important contractors on a large project and although it is not mentioned in any contract documents nor indeed has any strictly contractual value, it is nevertheless worth considering in these pages. It is known as the Project Proposal and it provides detailed design information on the complete permanent facilities or plant, including requirements applicable to mechanical features and construction materials.

Let us assume that a major operator has decided to build a new plant, and has issued a Project Proposal. The introduction will include a general description of the facility and of the planned engineering and construction work to be carried out. The type and quantity of feedstock will be mentioned, together with the major products expected. The location of the various production areas to be constructed will be explained, as will the overall master schedule, milestones, and proposed completion dates. Flow diagrams, storage and terminal facilities, utilities, including electric power distribution, process water, and gases, will receive mention together with general support facilities, administration buildings, roads and fencing, location of heliports, etc.

The introduction will be followed by a detailed project description with plot plans and technical data of the facilities, area by area, including utilities, buildings, pipelines, instrumentation and control, data logging systems, and communication system. Unless the units are protected by patents, the details of which are not to be divulged, chemical processes will be explained, including the reasons for the employment of special materials in construction.

Not all contractors will be allowed access to the project proposal, and its issue will be on a strict "need to know" basis. There is no doubt of its interest and value, particularly to the mechanical contractor. Even to the smaller cogs in the system, the acquisition of a bootlegged copy gives the possessor some idea of what he is helping to create. The Project Proposal should not be confused with any document issued during the bid stage, either by the client or by the bidding contractor.

The Project Record Book

This book is a bound collection of drawings and data relating to the project which the contractor is required to furnish as a ready reference for all important equipment items installed in the work. It is intended to assist in the initial start-up of the plant and to aid client personnel in subsequent operation, maintenance, and inspection. It also provides information necessary for future checks of equipment performance or for planning of plant expansion or redesign. The book will be used as an additional reference together with the project drawings, "as built," and specifications and will contain other important information not found in these documents. These may be vendor drawings and data sheets, manufacturers' information sheets, and the like.

The Construction Control Plan

This manual is produced by the client's construction team for the guidance of its field staff in the day-to-day supervision of the contractor's work. Each contract in the project which involves substantial hands-on construction work will have a C.C.P. produced as part of the quality control system. The extent to which the contractor is made responsible for QA/QC control activities varies from project to project but generally there are three levels of control.

- The first level is the contractual requirement that the contractor is fully responsible for his own QA/QC activities until released from the contract.
- The second level is the responsibility of the client or management contractor's site team and is performed by site inspectors for each discipline.
- The third level is the responsibility of the project's QA/QC department which will verify and audit all construction control activities.

The manual will include the Scope of Work for the relevant contract, job descriptions and duties of all supervisory field staff, and samples and explanations of all forms used in the control process.

CHAPTER SIX

Contracts Management

THE CONTRACTS MANAGER

Contract administration as a separate department of an oil company or a large engineering contractor is a comparatively recent addition to the construction industry. Forty or fifty years ago, producing oil companies would arrange construction agreements with contractors through their lawyers and engineers for major works, but most small civil undertakings were often carried out by the companies' own construction divisions. Gradually over the years, the producers, particularly in the fertile area of the Middle East, embarked on a policy of local purchasing for their material and materiel requirements, followed by a cautious entry into the employment of local contractors for selected work. Not surprisingly, the first contracts administrator was usually someone from the local purchasing department.

A similar pattern was followed by the leading engineering contractors. They were directly engaged by the oil industry as a hands-on contractor. They employed their own labor forces in most disciplines of construction, only subcontracting out in areas where they did not have expertise or it was uneconomical to attempt the work themselves. The contractor's buyers were also required to arrange agreements with the subcontractors and even when separate contracts departments evolved, they became attached to the procurement sections as a matter of expediency rather than logic. Although many companies still have their contracts engineers report to the procurement manager, the two departments have little in common in the field compared to the close relationship between the contracts personnel and the cost/scheduling department. This is especially apparent in the day-to-day contacts between the managing contractor and the subcontractors when they meet to discuss contractual matters. The contracts manager would not wish to attend the meeting without the presence of a cost engineer or the scheduler, but only rarely would he require the attendance of a representative of the procurement department.

The title of Contracts Manager is not new to the construction or manufacturing industries, but the job description is not the same for both. In manufacturing, the Contracts Manager is concerned with sales to clients rather than the direct administration of a number of construction contracts that make up the project in hand.

The emergence of the Contracts Department in the oil, gas, and petrochemical industries brought about a similar development within the organizations of the larger contractors. They were invited by the oil companies to manage projects and supervise the work of other participating contractors, a role for which the client companies very often had insufficient personnel or even the expertise to perform. In the palmy days of the 1970s, the very largest engineering and construction contractors were narrowly focused on the management of superprojects on behalf of the oil companies. Some of these ventures employed more contracts personnel than the client's own contracts staff. Sometimes the contract documents were written on the client's paper, but on many projects the managing contractor's procedures were used, and the contracts were issued by him with the client maintaining a fairly passive surveillance. Most contracts managers and contracts engineers make their way to the department through various engineering disciplines within the parent organization. In times of high unemployment in the field, many are attracted by the comforting thought that the contracts man is one of the first on a new project. This is because he must issue invitations to bid and produce and execute contracts before construction begins. The contracts manager is also certainly one of the last to leave a job, owing to the need to settle claims and arrange final close-outs. Not all contracts managers have engineering backgrounds. Some have trained as lawyers and others have a business degree. A few have transferred from project management and as contracts managers have been requested by clients to double as construction supervisors. The nomenclatures of contracts engineer and contracts administrator are not normally meant to differentiate between those of a technical background and others of a commercial or legal calling. On most occasions, the title is decided by the client or by the upper echelons of management. In the head office of a large managing contractor in the United States, objection was raised to the practice of using the term "contracts engineer" when applied to personnel who were not academically qualified to use the title. Rather than segregate the staff, everyone below the rank of contracts manager was addressed as contracts administrator. Throughout this book, both titles are used without consciously distinguishing one discipline from another.

THE CONTRACTS ENGINEER

It is advantageous if the contracts engineer who is to administer the contract in the field compiles the bid package and sits on the bid evaluation team. This seldom happens in fact and the field contracts man often inherits the contract from the home office without the benefit of knowing its early history. The authority of the contracts engineer varies considerably from project to project. In some construction undertakings he is the contractor's primary contact, chairs every meeting with the contractor, and carries a great deal of clout. In others, he may be placed in a lower rank. Generally, he has the administrative management and control of all contracts assigned to him and will be accountable for communications, documentation (files maintenance), status reporting, pricing, payments, changes, backcharges and offsets, claims, and business practice.

Of all members of the client's construction team, the contracts engineer is the one especially governed by project procedures. Every step he takes through the paperwork jungle of contract variations is preordained. There is even a procedure for stepping out of procedure! Contractor personnel, coming into the sphere of oil company or petrochemical construction for the first time, find it difficult to understand the rigid adherence to procedure imposed by the client. The same applies to some of the client's own construction personnel. "Let's get on with the job and worry about the paperwork afterwards" is a cry often heard in the field. It is not often *repeated,* however, around the larger oil companies who make an issue of "compliance or else!"

THE CLIENT'S TEAM

The numbers and formation of the client's project team in the field will depend to a large extent on how the project will be managed. It has been mentioned (in Chapter 1) that there are a number of options available including the employment of a managing contractor. If this option is adopted, the managing contractor will field the following departments and disciplines:

- Construction supervision–all disciplines
- Quality control
- Safety and security
- Material management
- Cost/scheduling and estimating
- Administration and finance
- Contracts management

The field contracts manager will deploy his contracts engineers according to the size of each contract. The largest, usually the civils and mechanicals, probably will require the undivided attention of a contracts engineer each. The smaller and later activities such as electrical, instrumentation, insulation, painting, and so on, will be administered by the remainder of the contracts team with perhaps two or three contracts per man. The contracts manager, while not normally directly supervising any of the contractors, will be present at each progress meeting and other meetings and will keep a watchful eye on the daily proceedings of the overall project and its contract activities.

DISTRIBUTION OF THE CONTRACT

After the contract has been printed, bound, and executed by both parties, three issues are selected as the ORIGINAL, the DUPLICATE ORIGINAL, and a COPY. The ORIGINAL is sent to the Company's archives, the DUPLICATE ORIGINAL to the Contractor, and the COPY to the field contracts manager. Other copies are made available to certain disciplines on a need to know basis but several unpriced copies may be distributed to field construction personnel. It behooves all recipients to read and understand the language of the contract. The contracts engineer must know exactly what the contract requires of the client and of the contractor and should insist upon compliance with the terms and conditions.

In the field, it is important that priced copies of the contract are kept under lock and key, particularly after site working hours. A sure sign of sloppy contracts administration is the desk top display of highly confidential documents left unprotected against dishonest attention. Even an old and closed out contract may be of value to a contractor bidding for a future but similar job. The contracts engineer should keep a spare copy of an unpriced contract for the use of field superintendents, and should make certain that all field staff should have a copy of the relevant scope of work.

EXECUTION

By the time the contract is signed and distributed, the successful contractor may have already started to mobilize his equipment and forces. Provision is usually made for this in the opening paragraph of the contract: "This Contract is made effective as of the _____ day of _____ 19 _____ ." This date could be a week or so before the contract is executed and is covered by the previously issued Letter of Intent—if such a step was taken to enable the contractor to prepare for the work.

MOBILIZATION

During mobilization, there is a mass of paperwork for the contractor to prepare and present to the client before he is allowed to start construction. For example, he would certainly not be allowed to start work before the client can see proof that his insurance obligations have been met, as required in the contract.

It is helpful if, before the preconstruction (kickoff) meeting, the contracts engineer prepares a list of documents required before and during the work. Here is a list taken from a typical refinery construction contract:

Description	When Required
Resumes and qualifications of all supervisory personnel:	Within 15 days of award
Appoint contractor's representative:	On award
Proof of insurance:	Before work starts
Obtain all necessary permits:	Before work starts
Submit working plan area:	Within 30 days of award
Welding procedures:	Within 60 days of award
Construction work plan:	Within 90 days of award
Hydrotest procedure:	90 days before testing
Daily reports:	Daily
Weekly reports:	Weekly
Monthly reports:	Usually by the 10th of the following month
Provision of Bank Guarantee (if allowed in lieu of retention):	Before first invoice

Description	When Required
Taxation reporting requirements:	As required
Contractor organization chart:	Before work starts
Update chart:	As required
Monthly safety reports:	Monthly
Baseline plan progress:	Weekly and monthly
Baseline plans:	Immediately after award
Manpower baseline:	Within 45 days of award
Cost baseline:	Within 45 days of award
Schedule narrative:	Monthly
Problem analysis report:	Monthly
Manpower reporting:	Monthly
Cost reporting:	As required
Cash flow forecast:	Monthly
Purchase order status:	Monthly
Material control system procedures:	Within 60 days of award
Over, short or damage report:	As required
Material close out report:	On completion of work
Material reconciliation report:	On completion of work
"As built" documentation:	On completion of work

Manuals to Be Prepared	When Required
Quality assurance:	Within 30 days of award
Safety:	Within 30 days of award
Planning-detailed schedules baselines:	Within 30 days of award
Progress measurement/reporting:	Within 30 days of award
Procurement:	Within 30 days of award
Material control:	Within 30 days of award
Work procedures for each phase of construction:	Before each phase starts

THE KICKOFF MEETING

During the administration of a multimillion dollar, two-year, mechanical contract, the contracts engineer will be required to conduct correspondence on contract matters, be totally responsible for all activities concerning contract documents, and chair or at least attend regular site meetings with the contractor. The very first meeting to take place will be the preconstruction or kickoff meeting. The first object of this meeting is to introduce the contractor to the client's construction team because, in theory at least, the contractor's staff have only dealt with the client's contracts department during the preaward stages.

Every discipline of the client's supervisory team will be present and will speak on the actions required and expected from the contractor.

The following is an example of a typical kickoff meeting agenda:

A. Opening
　1. Introduction
　2. Scope of Work
　3. Project philosophy
B. Organization
　1. Client organization and personnel
　2. Contractor organization and personnel
　3. Signature authority (Contract, correspondence)
C. Communications
　1. Oral
　2. Site instructions
　3. Correspondence
　4. Schedule for progress review meetings
D. Schedule
　1. Overall work plan and schedule
　2. Manpower curve
　3. Coordination with work schedule of others
　4. Contractor's schedule submittal
E. Materials
　1. Planning for contractor supplied materials
　2. Scheduled need dates for free issue materials
　3. Transport and offloading procedures
　4. Storage and/or requisitioning procedures
F. Client or contractor furnished facilities
　1. Utilities
　2. Accommodations or camp facilities
　3. Tools
　4. Shops
　5. Manpower
　6. Equipment
　7. Other services
G. Contractor's equipment planning
H. Laydown and work area planning
　1. Location
　2. Buildings or materials located in laydown area
　3. Conditions or restrictions on use

        4. Access
        5. Maintenance of access
I.   Labor relations
        1. Labor relations
        2. Situations with unions
        3. Site agreements
        4. Dispute handling
        5. Jurisdictional procedures
J.   Work rules
        1. Work week
        2. Overtime procedures
        3. Security provisions
        4. Other
K. Safety
        1. Regular safety meetings
        2. First aid and medical provisions
        3. Safety inspections
        4. Contractor safety program
        5. Safety committee establishment
L. Quality control program
        1. Contractor's quality control plan schedule
        2. Inspection and testing
        3. Unique requirements
        4. Independent lab
M. Measurement of units or progress
N. Invoice procedure
        1. Agreement of progress measurement
        2. Submittal of pro forma invoice
        3. Format of actual invoice
        4. Client internal invoice procedure
        5. Cutoff dates
        6. Method of deductions (backcharges, retention, etc.)
        7. Bank guarantee (if in lieu of retention)
O. Variations to the contract
        1. Work orders
        2. Change orders
        3. Amendments
P. Claims handling
Q. Reporting requirements
     (The previously listed documents and their due dates.)

R. Drawings and specifications
   1. Contractor requirements
S. Subcontractors
   1. Appointment
   2. Approval by client
T. Insurance
   1. Production of proof of insurance
   2. What type insurance required
U. Other business

INDEMNITY AND INSURANCES

The contractor is required to procure and maintain certain insurance covers before and during the performance of the work, including protection and indemnity insurance holding the client harmless from claims arising from death or injuries to third parties and/or damage to or loss of their property. The fiscal limit of this cover is set by the terms of the contract or is governed by the applicable laws of the country. The possibility of damage to client's property by the contractor, however, receives a special consideration. Construction of a new plant in the middle of an existing refinery, for example, increases the vulnerability of the contractor to penalties for damage far beyond the amount for which he would reasonably insure. Even if such cover was obtainable, the premium may render his bid uncompetitive. The client may ask the contractor to insure against serious damage to the permanent work or to adjacent property in a nominal sum, say $100,000. Thus, the contractor is not released entirely from all responsibility but is indemnified against liability for the remainder.

In certain circumstances where the risks are great, the client may go further and indemnify and hold the contractor harmless from all claims by the client for damage as follows:

(a) Any damage to or loss of client supplied materials during transport by the contractor
(b) Any damage to the work or facilities as described above exceeding $100,000 per occurrence prior to final acceptance
(c) Any damage as above after final acceptance to the extent that such claims are not covered by contractor's insurance
(d) Any damage to third party life and property not covered by contractor's insurance

(e) Any damage arising from the use of vehicles supplied by the client to contractor for use in the work
(f) Injuries or death of employees of the client caused by the contractor
(g) Any damage caused by loss of crude oil or other products due to contractor negligence
(h) Any damage caused by pollution

There are certain types of insurance cover that the contractor must assume at his expense. Oil company clients, because they own vast assets such as refineries in which the contractor may be working, will accept coverage of major risks either by self insurance or by worldwide brokerage. Consider a situation in which the anchors of a contractor's work barge damage a submarine pipeline, causing eruption of great quantities of oil. The premium for the contractor to insure against such a disaster may be greater than the contract is worth. Clients can always insure more cheaply in this direction.

However, the contractor will be required to insure against (for) Protection and Indemnity, General Third Party Liability, Personnel, and in some cases, Automobile Liability. He must produce proof of these covers before work is allowed to start.

Builders "All Risk" Insurance

This insurance, except as stated in the policy, will cover all property used in the performance of the work against all risk of physical loss or damage to machinery, materials, and other property which will become part of the work. Such cover shall apply, inter alia, while any such property is awaiting erection or installation, being erected or installed, and being tested.

General Third Party Liability Insurance

Both parties may take this cover, with the contractor's liability being limited to a small amount than that of the client.

Personnel Insurance (Workmans' Compensation)

This insurance covers the employees of the contractor for illness, personal injury, or accidental death to the full extent required by law applicable at any site and/or where the contractor's employees' contracts of employment are made.

Property Transit Insurance

This insurance will cover loss of or damage to the permanent works while they are being moved or transported.

Protection and Indemnity Insurance

This insurance covers the contractor's liability to hold the client harmless from and against all losses, expenses, and claims for death of or personal injury to third parties and/or damage to or loss of their property. The client will usually protect the contractor in a like manner.

PROGRESS REPORTING

From the beginning of every project, the contractor will be obliged to issue a monthly progress report to the client's representative. As a minimum, each progress report should include the following:

1. Highlights of significant accomplishments during the report period, expressed in relation to the total of work to be done in each category
2. Current status of the work. Project progress information shall be provided in the form of a monthly project update report showing actual progress versus scheduled progress for:
   (a) Detailed engineering
   (b) Materials commitment
   (c) Materials received at site
   (d) Field construction
   An explanation shall be given of deviations from the target schedules, their consequences, and corrective actions to be taken.
3. Problems encountered, together with actions taken to solve them
4. Highlights of significant work items anticipated to be completed in the succeeding month
5. Status of subcontracts
6. Contract price forecast for the project, including the cost effect of approved and pending changes to the work
7. Photographs of the site to indicate construction progress
8. Safety report

CONTRACTOR SCHEDULING REQUIREMENTS

One of the most difficult tasks facing the client's management team is to encourage the contractor in the timely production of schedule information. Under the terms of the contract, the contractor is required, as a minimum, to prepare an overall project schedule showing the start and completion dates for all major categories of engineering and field activities, vendor field delivery dates where applicable, and activities associated with subcontractors. Reporting requirements are usually related to the monthly progress report and will indicate:

- Separate progress curves showing the projected and actual progress for overall engineering, material commitments, and construction
- Estimated monthly manpower requirements to meet the progress previously projected, including the scheduled levels of designers, draftsmen, engineers, and personnel for each construction discipline and phase. Where approved subcontractors are involved, the manning schedule for these should be shown.

In addition, the contractor will be expected to maintain progress curves (planned versus actual performance) for each discipline. Progress will be measured by physical measurement of work completed.

THE CHANGE ORDER

Oil company construction contracts, with very few exceptions, include a clause which gives the client the right, at any time, to make variations in the character, quality, or quantity of the work that the client feels is necessary. Of course, having signed the contract, the contractor must carry out such work. In most client organizations, all such variations are made by means of a change order, work order, or an amendment.

The proposal for a change may be initiated by the contractor or by the client. If it is the contractor, he will usually draw the client's attention to work he has been required to carry out but which, in the contractor's opinion, is not covered in the contract scope of work and therefore deserves extra remuneration. The client will study this proposition and if he agrees, will issue a work order if the compensation can be covered by applying rates in the contract or if there are no applicable unit rates, he will initiate a negotiation for a change order. However, if the client is of the opinion that the work is not a deviation from the scope, he may reject the contractor's proposal and thereby place the contractor in a claim situation.

When the client proposes a change in the scope, it is most likely to be as a result of a design change, in which case the client will ask the contractor for a price for the extra work. At the same time, the client will make a fair price estimate of his own. If the two calculations are a match, say, within five to ten percent of each other, the change order may be executed. As with contracts, the client will endeavor to process a change order on a lump sum basis, but if this is not possible and a unit rate, day work, or reimbursable deal is unavoidable, some indication of the final cost should be given. This may be in the form of a notation to the effect that "THE VALUE OF THIS CHANGE ORDER IS NOT TO EXCEED $500,000" (known as a "not to exceed" change order).

A Change Order is issued to:

1. Revise the work, or specifications affecting the work
2. Perform repairs to deficient work (in reimbursable cost contracts only)
3. Correct or modify changed or unforeseen field conditions
4. Revise the work schedule, either at no cost if as a result of a force majeure situation, or at cost if as a result of a delay caused by the client
5. Authorize or approve standby time for client-caused delays

Before the contractor carries out any variation in the work, he is expected to advise the client of the effects, if any, on the contract price and the contract schedule. He is also not supposed to commence the extra work until he is in possession of a change order (if such is used) signed by the client's representative. One of the differences between an amendment and a change order is that the latter is an instruction, and the former an agreement in the same manner as the contract document.

Somewhere in every change order are the words, "The contractor is hereby ordered to proceed with the work described hereunder," or something to that effect. The inference is that the contractor has no choice but to carry out the dictates of a change order, even if he has to argue about how much he is to be paid for the work. Furthermore, the contract usually states that if the contractor should start the extra work without a signed change order, he may not get paid at all. This condition has led to many an altercation in the field, particularly when oral instructions have been given to the contractor to carry out work required urgently, but no written instructions have been issued. "After the fact" change orders have been known to bounce back, on presentation for signature.

Of course, there are nonroutine situations in which it would be very expensive to delay operations until a change order is fully executed. Consider, for example, a pipe-laying barge costing the client about $300,000 a day which faces a substantial change in scope far beyond the signature authority of the client's representative on board and requiring approval by a vice president. This happens, naturally, during the weekend when the VP is playing golf onshore and the laybarge is 150 miles offshore. If the barge is operating 24 hours a day, seven days a week, there has to be an operations room onshore with a manned telephone and telex facilities. The barge telexes the operations room with full details of the proposed change, someone goes out to the golf course, waits until the VP hits one of his better shots, presents the telexes, gets the go-ahead, and informs the barge accordingly. The change order may then be executed the following day. Most companies will still want that Vice President's authority, though, *before* the fact, even if it has to be scribbled on the bottom of a telex form.

One client, finding that delay over the negotiation of change order prices and subsequent delayed payment was seriously disrupting a large fabrication contract, decided to introduce a clause into the contract that obliged the contractor to make an immediate start on the work even before agreement on price. However, the company agreed to pay the price asked by the contractor, provided the details were fully documented. In the event of a serious dispute, the matter was to be referred to arbitration. If the company requested that the contractor should perform work which, in the contractor's opinion, was not in his scope of work according to the contract, the contractor would then ask for a change order and the client would issue one. If the company considered that the work was indeed part of the scope, this would be duly recorded on the change order. Now here is the interesting part: if the client, within six months after the issue of a disputed change order, has not announced its intention if appealing to arbitration, then it is considered that the matter is no longer in dispute! This novel arrangement was short-lived.

A more conservative experiment was carried out in the Middle East a few years ago in which a fairly large lump sum contract was awarded with the proviso that no change proposal would be presented by the contractor for work below the value of $3,000. In other words, the contractor included in his bid a sufficient amount to accommodate small packets of work extra to the scope and only requested a variation order when the work to be done cost more than $3,000. This arrangement worked very well and there is no record of the contractor being out of pocket. There were, of course, safeguards to discourage the contractor from saving up the smaller jobs until he reached the magic figure.

The old established companies in the oil industry are generally absolute sticklers for obedience to the rule that change orders must be executed before work may commence, even if there is a deadlock over the price and a "not to exceed" order must be issued. Where clients and their procedures are comparative newcomers to the game, the rules seem to be relaxed to the extent that "after the fact" change orders are accepted for signature without must fuss. Such laxity is not recommended and the client who condones such practices must eventually squander a good part of the project budget unnecessarily and encourage a "rubber stamp" attitude from the site team towards their signature authorities.

Change orders may be used in most extra work situations provided they fall within the general scope of the contracted work. Most change orders contain the following information:

1. Who initiated the C.O.? Contractor or client?
2. Type of payment agreed
    (a) Lump sum
    (b) Dayrate
    (c) Unit rate
    (d) Reimbursable
3. Backup information and correspondence leading to the change
4. Instructions to contractor regarding the work
5. Reason for the change
6. Cost, including cost codes, etc.
7. Start and completion dates for the work
8. Details of escalation, if any
9. Currency in which contractor is to be paid
10. Whether the price stated is fixed, an estimate, or a "not to exceed" value
11. Effect on contractor's baseline schedule, if any
12. Whether C.O. represents an increase in the contract price or a decrease (negative change)

Contracts should include, wherever possible, the unit price and dayrate schedules for pricing attachments associated with change orders. The simplest type of change order to issue is one which contains previously agreed prices from the contract for specific tasks. Failing this, valuation of change orders will have to be made by pro-rating the sums, rates, or prices contained in the contract, by estimating the manhours and materials to be expended and pricing in accordance with known rates, or by negotiation with the contractor. Whatever the method used, it is desirable (from the client's view) that a firm lump sum for the work emerges as the end result.

The change order should give some indication of how much the client may expect to pay, even if the amount stated in the change order is only an estimate. No client should allow a change order to remain open in this respect. If the work is expected to be ongoing, it is preferable to estimate a likely sum, add a percentage for contingency and apply a "not to exceed" figure on the change order. This will mean that the contractor cannot go past that amount or even approach it without alerting the client. There is no requirement for a "not to exceed" change order to be reissued should the value of the work performed be less than the NTE value. If the work looks as though it will be greater than the NTE amount before completion, it will be necessary to issue another change order.

The Negative Change Order

From time to time, an item in the scope of work may be deleted by mutual agreement between the client and the contractor. Sometimes this is because of a design change or perhaps simply because the task is no longer necessary. To avoid payment being made for work not carried out, especially in a lump sum contract, the client may issue a negative change order. The amount to be deducted will be negotiated in the same manner as a regular change order or interpolated from existing rates in the contract. In this situation, however, the contractor will be debited with the amount against the contract price. The negative change order is issued on a regular change order form but overstamped "NEGATIVE CHANGE ORDER."

The Change Order Closeout Form

This is a simple one-page form signed off by the contractor and client confirming that the work described in each change order has been completed and the full amount due has been paid. At first introduction, this seems like just another form dreamed up by some Head Office drone but, in fact, it is one of the better ideas in contract form production. It is particularly valuable in the closeout of "not to exceed" change orders where the final amount paid to the contractor is less than the NTE estimate on the change order. The change order closeout form signals the difference and assists in the reconciliation of the final contract price.

The Work Order

Within certain organizations, the Work Order is another name for a change order and is used in exactly the same way and in the same situations. However, the preferred use of the work order relates to circumstances in which the extra work to be performed by the contractor is within the general scope of the contract and the cost of such work is clearly described in contract unit rates or in any other predetermined price structure. This eliminates the necessity for negotiation. In this application, the work order does not replace the change order, but facilitates the expeditious performance of extra work by application of prices already negotiated.

THE AMENDMENT

Since the use of contract variation documents is developed in different ways from one company to the next, there is no hard-and-fast rule to be quoted for amendments, change orders, and work orders. Some organizations use amendments for all contract changes, work orders only for extra work against unit rates already established in the contract, and do not process change orders. Others use change orders for practically every contract change and will only issue an amendment if an alteration is required to the Articles. However, a study of major oil company procedures and those of managing contractors shows that change orders are usually employed to change the terms of the scope of work and for extra work for which there are no prices in the contract. Work orders are used when extra work is to be carried out but pricing may be obtained from the contract. Work orders may also be used for "service" jobs of comparatively low monetary value (e.g., an electrician's time for hooking up a power line to the client's site offices). Both work orders and change orders, when properly executed, are perfectly legal as far as the contract is concerned. The amendment, however, is a more formal document requiring different language and being signed, preferably, by the original signatories of the contract or, at the very least, by company officers of equal rank to the first signatories. For these reasons, the use of the amendment tends toward significant changes requiring more formal treatment. These may be:

- To change the terms of the Articles of Agreement
- To alter terms in parts of the contract other than the scope of work
- To suspend or terminate the work
- To alter the original contract price or unit rates
- To settle claims
- To add or delete substantial sums related to the scope, but which need more formal language not found in change orders
- To enable payments to be made to a party not specifically named in the contract (e.g., in a joint venture, when one of the partners wishes payment to be made over to the other partner rather than to the joint venture)
- Where there is a change of company title (of the client or of the contractor)
- To change the baseline master schedule, milestone dates, or completion date
- When the total cumulative value of change orders has reached an amount equal to 10% to 15% of the original contract price (In some organizations this is a safety measure to curb the excessive issue of change orders within the approval of the hierarchy.)
- Any other major modifications not within the general scope of the contracted work

The main structure of the amendment is in three sections and explains:

1. Amendments
   How and where the contract must be modified
2. Pricing summary
   What effect the modification has on the contract price
3. Status of contract
   The execution

An amendment may contain more than one change to the contract, including unrelated changes. Some systems involve the collection of a month's issue of work orders which are then incorporated into an amendment. This practice is not recommended. The beginning and the end of the amendment are couched in legal jargon, but the person preparing the technical and commercial parts should use simple language so that both sides may understand exactly what is being done.

THE SHORT FORM CONTRACT (SFC)

The SFC is a useful device for contracting work of a simple and nonrecurring nature with a limited value. The dollar limit of the Short Form Contract will vary from client to client, but usually will not exceed $50,000. The following rules govern the issue of a Short Form Contract:

1. Lump sum price or unit rates only will be used.
2. No progress payments will be allowed. Payment of the total value of the SFC will be made at the end of the work.
3. Work completion should be within six months of the date of commencement. Should unscheduled events occur to extend the work beyond the six-month period, the SFC will be terminated and the contractor paid off. If necessary, a new SFC will be negotiated to complete the work.
4. No variations orders or amendments will be made to a SFC. Should changes occur in the scope or work that were unanticipated at the time of issue of the SFC, the existing SFC shall be canceled and a new contract issued incorporating the necessary changes.
5. Bid procedure shall be similar to regular contract bid procedure excepting that selective bidding may be employed with a minimum of four bidders participating.
6. A company estimate shall be prepared for the SFC work and shall be kept in a sealed envelope until bid opening.
7. The SFC Articles of Agreement are not so comprehensive as the standard format.

Note:

Examples of Amendment, Change Order, Work Order, and Short Form Contract forms may be found in the Appendixes.

COMPLETION AND ACCEPTANCE

In considering the division of responsibilities between client and contractor for activities that occur during the completion period, it is necessary to regard the approach to completion as having three separate phases:

1. Substantial completion
2. Mechanical completion
3. Commissioning and "ready for start-up"

Unfortunately, most construction contracts used in the oil, gas, and petrochemical industries are less than crystal clear in the definitions of these stages and some contract documents fail to mention them entirely. The following words could be part of a typical refinery extension contract between client and contractor for the mechanical portion of the work:

> "When in Company Representative's opinion, the Work has been completed and has satisfactorily passed all tests that may be prescribed by this Contract, and Contractor has provided all data necessary for the operation of the Permanent Work, Contractor shall be entitled by written notice to Company's Representative to request a Completion Certificate. Company's Representative shall thereupon within a reasonable time of receipt of such notice from Contractor either:
>
> (a) issue such Completion Certificate stating that the Work is complete and has been satisfactorily performed, or
> (b) notify Contractor in writing of any work that remains to be performed before the Work shall have been completed in accordance with the provisions hereof.
>
> On receipt of such notice as mentioned in (b) above Contractor shall forthwith proceed to perform all outstanding Work prior to making a further request for Completion Certificate."

However, the contractor may not have had a hand in the design of the facility and almost certainly will not be responsible for commissioning and start-up, although he may leave a few of his key personnel to assist in this effort. The client's operations division, which will eventually control the plant, will want to reduce the number of contractor personnel to a level that will not interfere with safe operation. Therefore, one may safely accept the premise that, contractually, the contractor will be completely removed from responsibility involving commissioning and start-up and will be aiming at mechanical completion or possibly substantial completion. The Contract continues as follows:

> "Contractor warrants through the Guarantee Period that the Permanent Works shall conform to the final Drawings and Specifications and that the Permanent Works will be new and suitable for the purpose and use for which they are intended and not defective.

The Guarantee Period shall commence at the issuance of the Completion Certificate and expire 36 months thereafter or 12 months after the Permanent Works have come into commercial operation whichever period shall first expire."

The contractor can only guarantee work that he has performed. Unless they are his subcontractors, he cannot vouch for the work of the electrical, instrumentation, insulation, and civil contractors who will be required to sign similar and individual guarantees. It follows, therefore, that when the start-up button is pushed and the plant does not work or disappears in a cloud of smoke, the mechanical contractor is not necessarily to blame and before that stage and when certain conditions are satisfied, he is entitled to an honorable completion certificate. It is also possible that our contractor has got as far as he can towards mechanical completion but through circumstances beyond his control, such as nondelivery of free issue material or delay caused by another contractor, he cannot complete the scope or work. The contract sample given above does not mention mechanical completion or the steps to be taken when the contractor is prevented from achieving completion through no fault of his own.

Mechanical Completion

The work shall be considered to have reached mechanical completion when the permanent works described in the contract scope of work, or a portion or portions thereof, have been mechanically and structurally put in a tight and clean condition, and otherwise constructed as provided in the contract. All deficiencies, including those which could prevent or delay safe and orderly prestart-up or start-up procedures by Company, or timely achievement or permanent works operation at the conditions specified in the Contract, to the extent that such deficiencies may be determined, must have been corrected. As the contractor moves towards mechanical completion or an acceptable degree of substantial completion, the client inspectorate will present him with a checklist or punchlist of items requiring attention before a completion certificate may be issued. These items should be of a fairly minor nature since, in theory at any rate, the contractor has declared that the work is complete. The failure of the contractor to effect a thorough clean up of workmens' debris around the plant is a common item on the punchlist.

Substantial Completion

The construction of the permanent work may have reached a stage where only a few punchlist items remain, but the client considers that this work should not delay the issuance of a completion certificate for the purposes of bonus or progress payments to the contractor. It is also possible that such a stage has been reached in the work but for a number of reasons outside the contractor's control, he is unable to proceed further. In these events, the client may, by agreement with the contractor, issue a notice of acceptance based on substantial completion. The definition of substantial completion, in this case, is:

"Substantial completion shall be considered to have been achieved notwithstanding that; some amount of work, such as hydrotesting, minor repairs to coating and/or insulation, finish painting, removal of temporary facilities, or general cleanup remains to be done by Contractor, or some deficiency not affecting operation requires correction."

STANDARDIZATION

Favorite words throughout these pages seem to be "usually" and "varied" and such, when applied to clients' methods and procedures, since no two oil company contracts are alike. In the civil construction field of buildings, bridges, harbors, docks, and dams, there is an element of standardization of contracts through such bodies as the International Federation of Consulting Engineers, the American Institute of Architects, The Associated General Contractors of America, and domestic organizations in certain individual countries (e.g., I.C.E. of the United Kingdom). Where such conditions of contracts have been published, they have previously been approved by other institutions and contractors' organizations. In the oil, gas, and petrochemical industries, the contract document formats are produced without the benefit of outside consultation, least of all from the contractor.

We may well ask if the oil companies and the petrochemical industry worldwide would benefit from standardization in their contract documents. One contract system based on the best of all systems would surely save a great deal of time, money, and paper, in the long run. We have seen already that most superprojects tend to engage outside help for one particular project. There would seem to be an advantage in having these temporary employees arrive already trained in a standard system instead of having to move through a learning curve in grappling with yet another client's methods. Contractors would know what they may or may not do. Claims should be fewer because there will be fewer ambiguities from one client to another. Most major companies have systems that over the years have cost thousands of dollars to produce. Some of the smaller companies and most of the foreign national companies have no particular system and rely on the managing contractors to supply one. This means: for every different managing contractor, another different system.

To use a universal system would mean that the majors may have to make expensive adjustments, but these would be offset by advantages. Companies could pool their experiences with the problems of past contracts, and when a superproject is planned, this information could be shared to the benefit of all. At present, when faced with a contractual problem clients and contractors can point only to past court cases. But, as we have seen, these cases form only a small part of disputes arising from contract claims. If central guidelines were laid down for everyone in the association, at least some rules could be followed. If these rules were known in advance, bidders on a contract would know what to expect. The days of contractors bidding low and making up on claims would be limited. The attraction for the larger, older, and more developed companies is that many of their own procedures would be used anyway. For the smaller companies and inexperienced nationals, the possibility of expensive claims would be partly eliminated. Of course, there may be objections from the contract managers of the major companies, based on the conviction that their procedures and systems are the best procurable. (But, since they probably have not really studied anyone else's, how can they be sure?) As we have seen, no single system is infallible. It has also been demonstrated that the best system is the one to which everyone adheres in unity. It follows, therefore, that an *internationally* agreed system, incorporating the best from all methods, must be the best of all!

The first step is the creation of a panel of experts to draw up a universally acceptable set of model contract administration procedures as a foundation for each future project. The oil, gas, and petrochemical companies worldwide will be asked to assist in the foundation of this endeavor and to accept and abide by the precepts of the constituted panel. It would be of considerable assistance if the major producers would make their own contracting manuals and procedures available to the panel, since the object is not to reinvent the wheel but to emerge with some form of standardization taken from the mass of existing documentation.

The initial task of the experts would be to develop a framework of headings which could be presented as follows:

Contract Management Policy

1. High standards of ethics and fairness are to be maintained in client/contractor relationships.
2. All contracts are to be in writing.
3. Every effort shall be made to realize execution of the contract documents prior to start of work but if this is not possible, adequate documentary cover must be effected before commencement.
4. Only technically and financially qualified contractors shall be invited to bid for project work, and competitive bidding shall be employed. Negotiated contracts shall only be considered in exceptional circumstances and where there is no alternative method available.
5. Lump sum contracting is to be the preferred practice.
6. Contractor performance claims will be examined promptly and resolved in a fair and reasonable manner. Every endeavor will be made to produce contract documents that are succinct and precise in intent to minimize ambiguities and to lessen the likelihood of disputes. Claims which cannot be resolved within the confines of Project Management may, as a last resort, be referred to arbitration. Details of arbitration procedure will be described in the contract Articles of Agreement. Should access to suitable arbitration prove impractical or uneconomical, the Articles may provide an alternative in the appointment of an expert acceptable to both parties, whose judgment in the matter would be final and binding.

The Contract

A standard oil company construction contract will consist of several parts sometimes called exhibits, schedules or even just parts, as follows:

Part I. Articles of Agreement: The permanent and general part of the contract

Part II. The Scope of Work

Part III. The Master Schedule (timetable for the work) with milestone dates and critical path

Part IV. The compensation part: Details of contractor's remuneration and in some cases, unit rates and Bill of Quantity

Part V. Details of client-furnished materials and materiel, if applicable

Part VI. List of specifications

Part VII. List of drawing numbers

Part VIII. The section that alters or deletes those Articles in Part I above which, by mutual agreement between the client and the contractor, are considered irrelevant or unnecessary.

There are many variations and permutations in the selection of contract format suitable for the work but broadly, contracts may be classified as follows:

- Lump sum
- Lump sum with unit prices
- Unit rate
- Cost reimbursable with fixed fee
- Cost reimbursable with percentage fee
- Day rate

These formats are listed in order of desirability (to the client) and there is no doubt that the lump sum contract is preferred, provided the Scope of Work is well defined, materials arrive on time, few design changes are anticipated, and engineering is well advanced before the issue of the request for quotation. The lump sum contract, given these conditions, also offers the best cost saving potential, is likely to be more competitive, and provides incentive to meet schedules, since additional time carries costs intended for the contractor's account. In a situation where three or four projects have been completed in the same area and each one is a replica of the others, the lump sum method would be an obvious choice for the next project, as most problems will have been overcome and design changes will be at a minimum. Unfortunately, such conditions are rarely encountered and bidders are not always confident that they can accurately estimate the cost of the work without including excessive contingencies or assuming inordinate risks. To a certain degree, this can be overcome by adopting a lump sum arrangement for the portions of the work the client can be reasonably sure of and using unit rates for the uncertain areas. In a large contract, this is not always a satisfactory method as the unit rate portion tends to become open-ended. When times are difficult and business is slack and competition is fierce, contractors are obliged to accept lump sum or even turnkey conditions. In boom times, contractors can afford to be more selective. In the prosperous days of 1979 and 1980, certain of the largest contractors in the United States would accept contracts of below a million dollars only as a "favor" to old and valued clients. Ten years later, they would asphalt your driveway if given the chance!

Reimbursable cost contracting can carry a fixed fee for the contractor or a fee based on a percentage of the ultimate cost. Clients consider cost plus jobs only when there is no alternative. This may be a situation in which little or no engineering is completed and the scope is so vague as to prevent bidders from giving a meaningful offer. Compensation to the contractor under reimbursable cost contracts generally consists of three elements:

1. Reimbursable costs
2. Fixed rates
3. Fee

The first two elements are intended to cover the contractor's cost of performing the work under the contract. The third element is intended to cover the contractor's general cost of doing business and his profits related to performing the work under the contract.

The term reimbursable costs means those costs for performing the work for which the contractor will receive direct reimbursement from the client as opposed to indirect reimbursement via fixed rates. Reimbursable costs exclude the cost of materials, services, and other items, which are provided for in the fixed rates. They also include the contractor's general cost of doing business and profit, which are provided for in the fee.

Fixed rates are practical arrangements to pay the contractor for certain costs that he will incur in connection with the work under the contract. Fixed rates are normally used for certain of the contractor's office costs, such as payroll burdens for personnel, departments' overhead, and computer and reproduction costs. Fixed rates may also be used for certain other costs such as construction equipment and mobilization and demobilization of employees.

Fixed rates should permit the contractor to recover costs, that will be incurred and paid in connection with the work under the contract. Reasonable estimating contingencies and escalation allowances are usually permitted. Fixed rates should not be used as profit generators;

the contractor's profit is expected to be included in the fee. The fee provides for the contractor's income, profit, and general overhead related to the contract. Most clients will not permit the contractor to allow for the following expenses:

- The costs of salaries and travel expenses of executive officers who are not directly assigned to perform work for the contract
- Interest on capital employed or on borrowed money
- General and administrative overhead costs relating to general company and all office operations
- Consulting services, unless specifically requested and related to the contract
- Income tax
- Costs for employee bonuses and profit sharing plans
- Employee severance costs
- Employee relocation and recruiting expenses

Reimbursable cost contracts mean considerable audit surveillance on the part of the client company, far more than would be necessary with a hard money contract. Every invoice for goods and services, for example, would have to be checked by the client, whereas on a lump sum contract the client is not particularly interested in how much the contractor pays for material, provided it is within specification. The same applies to subcontracting.

Day rates would normally be used where, for example, large floating equipment spreads are employed offshore and the work carries a great degree of risk.

Contract Documents

When new drawings and specifications are introduced, for example, in connection with extra work through a change order, these should be included in the wording of the order, e.g., "Contractor shall procure materials, fabricate, and erect one building in accordance with Drawing A-BC-123 and Specification D-E-456 which, by this reference, are hereby made part of this Contract." It is important that engineering specifications and drawings should contain only essential technical instruction to the contractor and particularly in the matter of specifications, no attempt should be made to turn the document into a sort of "mini contract" by adding words that more properly belong in the scope of work.

Regarding contractual precedence of specifications and drawings, some contracts will define the order of precedence but it is usually the practice to include the following words in the scope of work:

"Anything shown in the drawings and not shown in the specifications or shown in the specifications and not shown in the drawings shall be of like effect as if shown in both and shall not be considered to be a conflict."

Similarly, anything shown in the drawings or specifications but not mentioned in the scope or work is usually covered by a paragraph in the scope on the same lines. Generally, the only part of the contract documents that has precedence over the other parts is the Articles of Agreement, but this precedence must be specifically mentioned in the contract to have any effect.

Conclusion on Model Contracts and Procedure Standardization

It is again emphasized that the model contract and the client procedures found in the Appendixes and Chapter 5 of this book are only a framework on which the proposed standardization panel could build a universally acceptable work of reference for use in the oil, gas, and petrochemical industries. They are, by necessity, brief and incomplete. It was not the intention of this work to offer a comprehensive treatise on conditions of contract and procedures but merely to propose an idea.

The only way to realize an efficacious standard is to borrow all model contracts and procedures from the major operators and have the panel sift through them, eventually emerging with one set of references forged into uniformity. The fruits of this endeavor could be offered to colleges and universities for the education of future project engineers, contract administrators, and contractor personnel of all disciplines, and at the same time, made available to all presently engaged in the industry. The panel of experts selected for this task would presumably be drawn from the ranks of the petroleum and petrochemical industries and from the leading engineering contractors. The composition of the panel may alter with each part of the model contract. For example, it would be necessary to have legal assistance with the Articles but not perhaps with the technical content. An experienced Manager of Contracts would need to be a permanent member of the panel.

CONFLICT OF INTEREST

We now approach the delicate subject of conflict of interest, or plain fraud. In the construction industry context, conflict of interest means transference of money, personal services, credit, or any other item of value, whether made in expectation of favors or not, from a contractor to a client's employee or managing contractor's employee. Most companies allow gifts such as advertising giveaways in the form of pens, pencils, calendars, and the like, or the reasonable cost of business entertainment. Throughout the bid evaluation period and the subsequent administration of contracts in the field, the contracts engineer is subjected to more temptation in this direction than most of the other company employees. Very often, relaxed security within the client organization will encourage conflict of interest situations to arise. Sometimes even senior officials will pass information to a contractor believing that they are doing nothing that could be described as dishonest. The same officials will see no reason to hand back that case of scotch or that gold watch at Christmastime. However, a contractor does not normally hand out expensive gifts with no thought of getting something in return.

In the boom years of 1975 to 1980, a large oil producer in the Middle East lost millions of dollars through malpractice by contracts and procurement staff. It is also cause for concern in Europe and the USA. Most fraudulent actions are eventually detected, but in a superproject involving multibillion dollar contracts and unscrupulous contractor/client staff, it is difficult to close the net completely.

It is almost impossible for anyone on the client's staff to cheat and get paid for it without the connivance of the contractor. It takes two to finagle. The relationship between higher management of a reputable international contractor and the lowest rank of field contracts engineer is so remote that fraudulent manipulation of the contract is unlikely, but among smaller or indigenous contractors, the danger is ever present. For a contracts engineer to swindle his employer, it is necessary to have the cooperation of the contractor and in some areas this is not difficult to obtain. Indeed, the initial approach often comes from the contractor.

A contracts engineer who is not on the client's permanent staff is usually engaged for the duration of the project. In a superproject with a life of perhaps five years or more, personnel may be taken on for a determined period. In overseas service, this could be one year on single status or two years married status. A contracts engineer may see the opportunity to make a considerable sum over his salary in a short time. Usually there are no prosecutions if he is caught, just the next aircraft home.

Despite elaborate precautions and procedures to prevent fraud, an enterprising administrator can always find a way to cheat. Most of the tricks used are quite simple. One method is the substitution of a page or pages in the contract. It has been suggested that this can be thwarted by binding the contract so that the paper cannot be removed without detection and having both parties initial each page. This may work if an attempt is made to alter the contract *after* execution, but it is not much use if the alteration or substitution is done *before* signature. Most contracts are produced on word processors and the chances are good that the operator has no close knowledge of the subject and little interest. How many auditors check the new contract with the contractor's bid? Not many. The client's finance department pays contractors' invoices following verification by the field and after checking the contract, amendments, and change orders. It is therefore unlikely that anyone in the finance department will uncover a fraud perpetrated before contract execution. Clients could go a long way toward prevention of this type of swindle by having auditors compare the contract with the bid documents immediately after award, and also by checking the drawings and specifications against the scope of work to satisfy themselves that there have been no recent and unauthorized additions to the scope.

The most common deceit occurs in the processing of change orders and claims. The hard money or lump sum contract of any size that does not include a schedule of unit rates for future variations to the work extends an invitation to the contractor to make extra profit on change orders and claims. Where rates have to be negotiated, there is always a possibility that more will be paid out than is justified. Although the contracts engineer does not finally approve negotiations, he undoubtedly influences the end result.

The Diplomatic Brush-Off

A senior contracts manager involved in a large Middle East construction venture was approached daily by contractors offering a thousand dollars here or a couple thousand there to fix or use his influence in the matter of a spurious claim or change proposal. He would sit the contractor down in his office, close the door and say, "Look, I am not adverse to taking a bribe." The contractor's eyes would light up. "It is really a question of how much. You see, I would stand to lose a lot if this was discovered. To start with, there is my salary of about $100,000 a year. I reckon that I have about another ten years with the company, so I would stand to lose $1,000,000 if I get fired. Then there is my pension and my trust fund. Let's say about $500,000. So, I am quite prepared to do business with you, provided we start talking in the region of say, $2,000,000 as a minimum. So, what sort of a deal did you have in mind?" Then would follow the rapid exit of the contractor, who was thinking in terms of $1,000!

CHAPTER SEVEN

Claims

CONSTRUCTION CLAIMS

There are many excellent works on the subject of the law and construction that justifiably are in constant use by contractors, oil companies, and students of contracts management. The authors have a mainly legal background and include in their offerings case histories and legal decisions made in the law courts in connection with contractors and clients. It is generally accepted, however, that for every claim that reaches arbitration or the courts, there are numerous disputes that are settled between the parties involved without a lawyer in sight. No public announcement or record is made of these situations and it could be said that such events are of no concern to anyone outside the contractor/client conclave. Nevertheless, it would be a pity if the circumstances leading up to a contractor claim were not examined as a matter of interest and possibly for the future guidance of proponents from both sides. In the following pages, an attempt is made to bring out into the light, dust off, and inspect some past claims made on oil, gas, and petrochemical projects in various countries and against various clients.

In a claim situation, it is an advantage for the contractor to understand client reaction towards claims and to do this it is necessary to probe right back to the bidding stage and the emergence of the bid package to find out the most likely causes of claims in the first instance. For our purpose, a contractor claim may be defined as a demand for payment believed to be due but at least in the initial stages, denied by the client. If it is not disputed, then it is not a claim and becomes the subject of a change order. If only partly disputed, then the part not in dispute will be allowed and the balance will remain in a claim situation. When a claim of comparatively modest proportion is firmly rejected by the client, some contractors will capitulate, not wishing to take the matter any further (certainly not to the law). The contractor may fear loss of position on future bidders' lists, or consider that continuing the struggle is just not worth the effort.

It is worth noting that larger international contractors will pursue a claim with far more determination than smaller or perhaps indigenous contractors. It is also interesting to observe that the latter will usually do better on realization of claims when in joint venture with the former. Is this by virtue of superior preparation or presentation, or some anticipation in the early stages of the contract? It is probably a combination of all three, together with the staff to handle such claims. Is it possible that some contractors examine the bid package not only to offer the lowest bid, but also to expose the weak links in the embryonic contract's armor against claim invasion? The indication is that not only is it possible but all too prevalent, particularly in the present atmosphere of recession.

A CLAIM BY ANY OTHER NAME

Foremost among the claims champions are those contractors who, while they do not hesitate in the normal conduct of their business to call a spade a spade, are remarkably coy about calling a claim a claim. To them, all requests for more money are change proposals and only at the moment of final rejection should a change proposal be called a claim. There is, of course, an element of psychology in this approach and the aim is to get the client thinking along the lines of a case for contract variation, rather than claim rejection at the outset. As a result, even the most outrageous claim is presented in wide-eyed innocence on a change proposal form as if the change order will follow almost automatically.

An international civil contractor engaged in the construction of facilities worth $25 million closed out the job without once ever submitting a claim. He forwarded 216 change proposals though, only 58 of which were immediately accepted as bona fide design changes, extra work, and the like. The remainder, all claims really, were argued over and debated and in the end, he won a large proportion mainly by pretending that they were not claims to start with and by persisting until he had at least some concession from the client, worn down by tenacity rather than the merits of the submittals.

In an obscure way, the client is also nervous whenever a claim is mentioned. Often, although project procedures on change orders may run into several pages, the client procedure on claims, if there is one, is usually a modest affair. It is almost as if the client does not want to dwell on the subject too long.

It should be mentioned that we are discussing contract related claims, which are the responsibility of the client's contracts department. Claims for loss or damage to property, personal injury, or death would be dealt with by the insurance department.

THE BREEDING GROUND

Claims germinate throughout the bid preparation stage and after execution of the contract. Even in the first weeks of project planning, some action or inaction may create a future claim situation. In chronological order, potential claims may arise through:

1. Faulty project planning
   - incomplete or incorrect design
   - inadequately defined scopes of work
   - limited time given for bid development
   - insufficient site investigation
2. Poor material and materiel planning
   - delayed ordering of free issue material
   - vague or conflicting specifications
   - no checks on lead time and availability
3. Poor choice of contract format
4. Errors and omissions in bid package carried over unchecked into executed contract
5. Acceptance of ultra-low bidder
6. Shortcuts in contract preparation
7. Failure to highlight scope requirements necessary for the satisfactory performance of the work
8. Perfunctory examination of contract documents by successful bidder in haste to effect execution
9. Failure by client to introduce strict controls and contract compliance on reimbursable cost contractor

CLIENT OBLIGATIONS TO THE CONTRACTOR

A typical oil industry construction contract does not cover all the obligations of either party even in the most comprehensive Articles of Agreement. The contract is mostly silent on the subject but past legal actions have indicated that there are certain obligations due not only from the contractor but from the client also. From contract award and the start of the work, the client has an obligation not to interfere with the contractor's operations or to delay him in the normal course of his work activities. The contractor may be in a claim situation when the client has:

- Given incorrect information which misleads the contractor in the performance of his work
- Not passed on information to the contractor necessary to the satisfactory performance of his work
- Instructed the contractor to carry out tasks in the scope of work using a particular method when the contract does not specify any particular method; or assumed direction of parts of the work which are properly the responsibility of the contractor without good reasons covered by the contract documents
- Failed to have work performed by others which is scheduled to precede the work of the contractor
- Rearranged the contractor's schedule in a way that obliges him to carry out work in adverse conditions (e.g., causes work which the contractor intended to carry out in the summer to be delayed until winter)
- Failed to inspect the contractor's work within a reasonable period
- Caused delay which could have been reasonably avoided

CLAIM REVIEW

When the contracts engineer receives a change proposal that he considers is not a justifiable prelude to a contract change order, he will disallow the whole or part and attempt to reach agreement with the contractor to withdraw or negotiate that part which is not in dispute. If the disputed portion still persists, the matter will be registered as a claim. Every client organization has levels of signature authority in the settlement of claims, from agreement by the site team in the field to the establishment of a claim review panel somewhere in the higher echelons. Most clients also maintain an appeal board which will convene if the claim review panel has finally rejected the contractor's claim and/or an agreement cannot be reached. The client will advise the contractor that the decision by the appeals board is final but if the contractor refuses to accept the final offer, the matter will be handled by the law department, either in arbitration or in court. Both sides are naturally reluctant to take this step.

A contractor may notice a certain anomaly in the contract shortly after award but will do nothing about it until it actually starts to hurt commercially. Most contractors start thinking about claims too late in the contract and will only claim months after the problem first appears, when it finally seems to be costing too much money. By the time they get around to it, records have been lost, key personnel have moved off the site, and their case has been weakened by lack of documentation. It may come as a surprise to some contractors and as a painful revelation to clients, but the time to start looking for potential claims is immediately upon receipt of the invitation to bid!

Many contractors will not pursue a claim rejected in the early stages on the ground that such action may jeopardize their chances of obtaining future work from that particular client. The indications are that this fear is generally without foundation and clients are more likely to remove a contractor's name from future bidders' lists owing to indifferent performance or technical shortcomings rather than the propensity to make claims. At the closing out of each major contract, clients will prepare a performance report on the contractor and it is unlikely that an adverse report will be written only on the claims situation unless it can be shown that the contractor deliberately set out from the start to produce spurious claims. That is very difficult to prove and commit in writing. In a claim situation, it is likely that the client is able to field more researchers on the job than the contractor, with more time to go through the files to find ammunition for claim defense or counterattack. This is all the more reason why the contractor should start in as early as possible and build up each case with full supporting documentation. In the construction of his claim plan, the contractor should start at the bid response stage to refresh his memory on exactly what was promised at the time. The client's team most certainly will be doing just that. The importance of keeping daily logs and full backup information, including timesheets in support of standby, etc., cannot be too strongly emphasized. The client's staff is required to justify each variation to the contract and leave an auditable trail by way of detailed backup information in each change order file. The contractor, on the other hand, seldom has such inhibitions and his primary target is to get his money paid without deferring to auditors. Nevertheless, he needs the back-up and records to support and strengthen his claims.

In the oil construction industry, very few contractor claims reach arbitration or court settlement considering the number of contracts awarded. For this reason, few claims histories are released for posterity or as studies for the industry in law or in general. Oil companies are somewhat reluctant to release details of claims settled with contractors out of court, as there are sometimes skeletons in the closet which are best forgotten.

There are many excellent works produced by lawyers, who specialize in contract law, quoting judgments in cases that have appeared before the courts. These cases amount only to the tip of the iceberg when one considers all the cases settled out of court. It is not in the client's interest or that of the contractor's to reach court appearance until every conceivable avenue has been explored. It is a pity, of course, that details of such claims are not published for the benefit of all who wish to study the pitfalls of this contracting business and to learn from the mistakes of others. In many of our major industries and professions, conferences and seminars are held regularly to exchange ideas and information for the free use of all in the business. Industrial secrets and processes are not unwrapped in front of competitors, naturally, but useful updates on safety, security, and the like are discussed. Details of claims settlements, however, are seldom disclosed possibly because most claims are indirectly related to mistakes made in the preparation of the contract and few administrations are willing to admit to such errors.

CONTRACTOR TENACITY IN CLAIMS PROMOTION

A surprising number of claims seem to be presented without merit and even without foundation yet many of these result in victory for the contractor. Some old hands in the business might say, "We certainly would not pay those claims and the contractor would be told in no uncertain manner than he would be wasting his time even thinking about a claim." Faced with this response, the majority of contractors would withdraw the claim. But what of those who do not back off; those who keep pounding away and resubmit the claim at each progress meeting? They get paid. Why do they get paid? Because most clients will afford the benefit of the doubt to a contractor who claims that he did not include that particular item in his bid, especially if the client can persuade himself that there may have been an anomaly in the contract. However, the contractor must be tenacious in his determination to promote his case and be ever watchful for that chink in the contract's armor.

Following a large refinery expansion in South America, analysis was carried out on claims presented during the three years' duration of the work. Out of one hundred contracts, 85 were awarded to indigenous companies and the remainder to large international contractors. The total value of the construction was split about equally between the locals and the internationals. The number of performance claims successfully pursued by the international contractors far exceeded those of the South American companies. The local contractors would present fairly sound claims, but on the first rejection would withdraw and never be heard from again on the subject. The foreign contractors, on the other hand, lined up their claims teams and went to battle.

CHANGE ORDER AND CLAIMS MEETINGS

In field operations, the contractor will be obliged to meet with the client at regular intervals to participate in site progress meetings. The client will often include in the agenda a status report from the contractor on potential change orders and claims. The object of this requirement, particularly with respect to claims, is to have the contractor declare his current intentions in this direction. Some clients do not include this item as they believe that it encourages the contractor to think about claims long before they exist. However, the practice is being widely adopted. A declaration in the minutes to the effect that a contractor has no claim at that time may not prevent him producing one at some later date. However, these status reports give the client some warning and help to prevent the surprise presentation of a carefully constructed claim six months after the event, or just towards the end of the work.

The contracts engineer is not necessarily a lawyer (in some cases he is not even an engineer). He is not required to interpret the legal aspects of the contract or to form an opinion of the likely outcome of court action between the client and the contractor. This is properly the domain of the client's legal department. Nevertheless, he must define and defend the client's position on contractor claims, most of which will not be brought to the attention of the company lawyers. By holding regular change order meetings with the contractor and by generally keeping his finger on the pulse of contract activities, the contract engineer will sense the arrival of a claim long before it is formally submitted and will have prepared backup information and a daily log on the subject, opened up a file on the claim, and given it a number. When the claim arrives, probably as a change proposal, the contract engineer will be prepared for it and will have held preliminary discussions about it with the rest of the site team. It has been noticed that sympathy for the contractor's claim will broaden as it progresses up the management ladder. The clients' construction team closest to the work will tend to reject most claims at first sight but if the contractor persists and the claim reaches the attention of project management, a more benevolent attitude will usually prevail. Attitude is the key word here; contrary to popular misconception, the legal merits of such cases receive scant attention at the lower levels. In the absence of guidance from contract documents, the position taken by the site team is very often based on the conviction that the contractor is trying to get away with something to which he is not entitled. Often, a cry is heard to the effect that "We have paid this contractor enough already and we are not paying any more!" This sentiment is often expressed before the merits of the claim have been fully discussed. It must be admitted that the contractor's progress with a difficult claim is sometimes made easier by a chat between principals, i.e., the client's project director or vice president and his opposite number in the contractor's camp.

Contractor: Bring up the big guns as soon as possible when a sizeable claim is having a rough passage. Ask for an appointment at the highest level in the client organization and keep the lower echelon out of sight but handy for fact updating.

Client: Avoid these chummy meetings like the plague. Unless there is no escape, insist on claims being dealt with by those paid to handle them. The same applies to proposed meetings with contractors' lawyers.

CONTRACTOR'S CLAIMS PREPARATION AND PRESENTATION

A number of reasons have been forwarded as to why a claim succeeds or fails: attitude of management, priority of schedule, project budget status, etc. Any claim, whatever its merit will be difficult to promote if insufficient care is given to preparation and presentation. Confidence and optimism are the first ingredients, and that is why claims should be prepared as change proposals in the first instance as if the next step is automatically the change order or amendment. Contractors should abandon the CP approach only when the case is obviously hopeless.

It should be remembered that large oil or petrochemical industry construction contracts that are breaking new ground technically or planned for comparatively undeveloped parts of the world seldom if ever contain complete information on the nature of the work. This condition, coupled with an almost traditional haste to produce the bid package and the limited time allowed for submission of quotations, invariably results in errors and omissions in the contract. Some of these omissions may not be apparent in the bid stage even if the client's staff have all the time in the world to produce the bid documents. In a lump sum contract, the bidder is often expected to provide for most contingencies, foreseen or not. The contractor who deliberately bids low hoping to win the contract and make up the difference in claims deserves no assistance, but the honest bidder, driven to accept loose conditions by circumstances of economic recession, needs all the help he can get.

It may be thought that some of these arguments are biased in favor of the contractor. However, it must be suggested that the client has a partial remedy and that is to allow plenty of time for the preparation of the invitation to bid, to provide maximum design engineering, and review the package thoroughly before release to the bidders. Additionally, the client should always conduct a job explanation meeting and site visit. There is a further observation to be made in defense of the contractor. In civil construction works outside the oil industry, the client is usually represented by an engineer appointed by the client or owner who is at the same time an independent member of a professional body. In theory, at least, the engineer of integrity provides a buffer between the contractor and the client and will often persuade the client that a contractor's claim is fair and reasonable. In oil industry construction, the contractor is lined up against direct employees of the client or managing contractor who are there to look after the interests of the oil company. The passage of a relatively small claim, not worth bringing to arbitration or law, is entirely in the hands of the client's management.

Contractor claims should be prepared in close consultation with engineering and construction personnel. Preparation should not be left to someone on the contractor's staff who is far removed from the situation at the site or from where the claim originated. Provided the facts are sound, everything should be put into the claim that credulity will bear. It is better to retreat and concede a point or two than have nothing to concede. In the actual presentation to the client, the claim should be neatly typed and contained in a good quality cover. A sloppy presentation will get the attention it deserves.

The contractor should not put the blame for the client situation on the client's site team as if he were the essence of purity and innocence.

If he was wrong in certain areas, he should admit it. This may have the effect of defusing the site team's arguments and gaining sympathy from client project management. If all seems lost and the contractor is faced with the choice of arbitration or nothing (a choice he may not particularly relish), it is time to plead poverty, misunderstanding of the contract, different intention at time of bid and anything which will improve his chances of salvaging something in the way of a sympathy claim. It is surprising how often this works.

Claims should be presented as soon as possible and at the highest vantage point on the progress curve. Contractors should not wait until the last stages of the job when half of the key personnel have been demobilized and those remaining know little or nothing of the claim history. Do not wait until the schedule is met and the client has little to gain by being generous.

The curve shown in Figure 1 illustrates the comparative negotiating strengths of contractor and client over claims and change proposals throughout the life of the contract. During the bid period, the client has the advantage particularly when it dawns on the bidder that he may be in with a chance of success. Directly following contract award, the happy pair move into the honeymoon phase; all is sweetness and light and the contractor is anxious to please. At this time it is not unknown for the contractor to carry out little tasks of extra work for free. Later on in the contract, he will most certainly demand payment for similar work. The honeymoon period, alas, is soon over. The client starts worrying about the schedule, perhaps accusing the contractor of not getting on with the job fast enough, or of insufficient manpower and/or equipment. There are even times when the client contemplates dissolution of the marriage but is dissuaded by the realization that he will have to go out to bid again (more expense and time) and that probably he will not be much better off with a new partner.

Adherence to the schedule becomes important. It is important to the contractor too, but to a lesser degree. The client's curve now moves to a position of weakness and the contractor is in the bargaining saddle. However, as a contract moves towards its close, the client begins to regain negotiating supremacy and when, say 95% completion is achieved and only minor tasks and check list items remain, contractor claims are viewed with less sympathy. The lesson shown by Figure 1 serves again to remind contractors that keeping claims back until the last stages when the job is almost finished is not recommended.

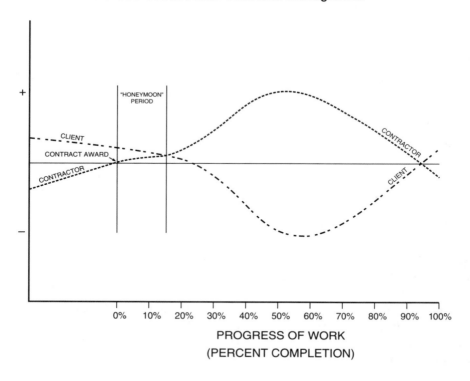

Figure 1. Comparative negotiating strengths of contractor and client over the life of the contract

---

ARBITRATION

On the face of it, arbitration seems simple enough to arrange. In broad principle, each side chooses one person to be an arbitrator and the two arbitrators select a third to sit in judgment. Most Western countries have an arbitration institute and will arrange to provide arbitrators who are experts in the construction business or in whatever field the dispute lies, and will fix the venue and the procedures for the hearings. The fees of the arbitrators will be determined by the time spent on the case, the amount in dispute, and the complexity of the matter. There will also be costs for travel and lodging, secretarial assistance, mail and telephone, telex, telefax, etc. In addition to all this, a fixed amount for administration costs will be due from the claimant to the arbitration institute. The administration costs will be based on the amount of the claim. In practice, the problem with arbitration is the time it takes, not only to get started, but to reach some sort of decision. It has been known for the hearings to be spread out over a year or so.

NEGOTIATION

The contractor has finally documented and presented his claim. The client's team has also prepared the case history, starting to keep records and creating a dossier perhaps long before the claim is formally tabled. There is, of course, disagreement over time or money or both. This dispute has reached an apparent deadlock and meetings have been arranged to reach some sort of agreement and avoid arbitration.

For the client, *control* is the key to a satisfactory settlement. It is, after all, the client's money the contractor is seeking and the client is entitled to call the tune. The tune, in this situation is the method of conducting the negotiation meetings. The venue for the meetings should be the client's premises or at least in a neutral place. Client control loses some of its strength in the contractor's conference room. If the claim is sizeable, it is unlikely that a settlement will be reached in the first sitting. At the initial meeting, therefore, the client should allow the contractor to speak first, on the premise that the contractor by presenting the claim is in fact, requesting the meeting. It is not recommended that full minutes be taken at the initial meetings. Only records of milestones in the progress of the discussion or salient headings should be noted. It is definitely not in the client's interest to have minutes recorded by the contractor at any of the negotiating meetings. At certain times in the proceedings the meeting may be interrupted to enable private discussions to take place by either party. When this happens, it should be the client who requests the contractor to leave the room for a short time to allow these sessions to take place. Hence the necessity for the meetings to be conducted on client territory and for the client to establish authority at this time and throughout the hearing. If the meetings were held on contractor premises, it would be difficult indeed for the client to stand out in the corridor to commune with his team and his authority would suffer if this was even considered. Both sides may benefit from the following advice.

Before the meeting:

- Be armed with full backup records and documentation
- Appoint a spokesperson who may call upon each member of the team to provide facts and figures as required
- Determine the limits of authority for the team
- Keep a fallback level of flexibility
- Prepare a plan of action for all foreseeable contingencies
- Arrange a rehearsal meeting with your own team

During the meeting:

- Be courteous throughout
- Keep cool and calm
- Schedule breaks during the meeting
- Maintain a pleasant atmosphere without getting familiar or flippant with the other side

CLAIMS AFTER CONTRACT COMPLETION

The Certification and Release Agreement signed by both parties at the completion of the work is accompanied by the client's final payment to the contractor. This document affirms that the client has paid in full for all labor, services, materials, and other items furnished by the contractor in connection with the performance of the contract. The contractor agrees to indemnify and hold the client harmless from and against any and all claims, liability, and loss which they or either of them may incur as a result of a breach of the agreement. Nevertheless, there have been situations where a contractor has submitted a claim to the client even after signing the release agreement and receiving the final payment. When this occurs, the client will naturally advise the contractor that, by signing the release, he has surrendered his right to claim under the contract. If the contractor feels that he has a good chance of persuading the law courts that his claim is worthy of attention despite the release agreement, he may pursue the matter with the client for some ex gratia payment. The contractor needs plenty of patience and determination since the work has been completed, presumably satisfactorily, and the client is probably not in too much of a hurry to hand over any more cash. More often than not, a successful claim of this nature originates from a third party. This can be a supplier to the contractor for materials used in the completed project, or a subcontractor for services rendered but not previously invoiced and not protected by mechanic's lien statutes. If the client believes that there was a genuine oversight he may be persuaded to pay up. On the other hand, he may not!

SUBSTANTIAL CLAIMS

Very large claims, in excess of say, $50 million, are seldom settled in the field. The contractor will almost certainly engage the services of a claim consultancy to prepare and present his case to the client, to a panel of experts convened under the terms of the contract, or perhaps to arbitration. On the arrival of the consultant's team, a familiar pattern involves the division of the group into writers and leaders. The first task of the writers is to create a library of copies of all correspondence between the client and the contractor and all documents relevant to the claim. Assuming that the team has studied the contract, the next step is to read through all the correspondence, and at the same time conduct discussions on the claim's history with the contractor's project staff. Large construction claims are segmented, that is, divisible by the nature of the work and cost but perhaps joined to the others in forming a global claim for delay and disruption. Each writer would start on one or more of the segments. One may create a dossier on the civil portion of the claim, another researches and gathers information on the mechanical part, and a third may undertake the electrical and instrumentation section. A fourth writer may work on the miscellaneous remainder portions embracing insulation, painting, etc.

During the compilation of his dossier or "box," the writer needs to research and read through every document and correspondence relevant to his subject. An important aid in this endeavor is the storage and retrieval system available on computer software packages in most large establishments. All the correspondence between contractor and client is transferred within the database to the main computer and the users may track documents in accordance with the selected criteria. The writers, in collecting material for their separate boxes, will prepare a historical narrative on each part of the claim. They begin with contractual references and a description of exactly what the contractor promised to do according to the scope of work and other relevant parts of the contract. This is followed by a chronological description of events leading up to the claim, backed up by references to correspondence, change order proposals, and other supporting documents. At this level, there will be no mention of cost, loss of profit, or any aspect of finance. Only a well documented history of the claim showing how the contractor was placed in a claim situation is presented. The other half of the consultant's team, the leaders, will examine the material in the boxes and from this information will prepare the total claim submission. They also put a price on the collection, plus an amount for delay, disruption, inefficiency, etc.

"**Delay and disruption**" is the term used to describe the indirect, secondary impacts (such as added costs) on a project that result from the direct impacts associated with unscheduled events and conditions. Delay and disruption claims are complex and more difficult to promote than claims against direct impacts. Their promotion is made harder by the realization that there is no universally accepted method of assessing disruption costs and most of the methods used are contentious. A growing number of claims consultancies offer a computer simulation of delay and disruption as an aid to preparation of claims submittals. The aim is to recreate, as closely as possible, the actual man-hours expended on the project by simulating both the original and additional work that had to be done, and the disruptive effect of changes that impacted the project. The stages in the technique for calculating delay and disruption impacts are as follows:

1. Preparation of an "as built" model
2. Identification of disrupting events and changes in relation to the as built model
3. Extraction of the disrupting events from the as built model to arrive at a "no disruptions" model
4. Calculation of the total costs of disruption by subtraction of the total man-hours/cost represented by the no disruption model from the total of the as built model.

The as built model includes the original project program plus all the changes and delays along the adjusted schedules that have been derived from correspondence, other agreed documents, and contractor staff recollections. As the model proceeds with its simulation, it absorbs input of data relating to the direct impacts only, and then simulates the indirect effect of each change.

When there is a series of interacting, disrupting events on a project, it is not possible to apportion additional costs to the individual cost generating events. The advantage of the computer simulation is that it can efficiently arrive at the total disrupting effect.

It is essential that the as built model is based upon factual records which can be jointly agreed between client and contractor. No assumptions or assessments should be necessary in the construction of the as built model. Identification of the causes of delay and disruption is based preferably upon agreed records but in any event, documentation and backup information must be comprehensive and unequivocal. The extraction of the disrupting events from the as built model establishes the number of man-hours that would have been expended if the changes that impacted the project had not occurred. The difference between the no disruption model and the as built model is the assessment of the additional man-hours and cost caused by the disruption. This result supports the claim for delay and disruption. The results of computer simulation of delay and disruption claims on their own probably are not admissible in a court of law or in arbitration. Indeed, many consultants would not introduce the method as a panacea or intend it to be self-supporting. Generally, the models offered simulate only one component of project costs: the expenditure of labor hours by discipline. Nevertheless, such systems have been used successfully to support claims prepared in the traditional manner.

RECAPITULATION

In these days of slogans, catchwords, and other aids to the stimulation of commercial efforts, it may be appropriate to offer three key words of advice and encouragement to the contractor caught in a claims situation.

These are:

1. Preparation of the facts and figures
2. Presentation in understandable and businesslike form
3. Perseverance, or single-minded tenacity in the promotion and realization of the claimed amount

Case Histories

Caused By: Vague Contract Language

A contract was awarded in which the client was to have supplied some free issue material and the contractor was also responsible for the provision of certain material. The contract was fairly clear on the division until it ran into the word "provided." Part of the scope of work involved the setting of large anchor flanges into concrete blocks. The flanges were fabricated by others and client supplied as free issue material. The contract said that the flanges will be provided, meaning supplied. Elsewhere in the specifications, the engineers had unfortunately used the word in a different context. They said, "The thrust bearing flanges shall be provided (meaning "prepared") with an asphalt coating." What they meant to say was, "The client will provide the flanges but the contractor will provide and apply the coating." The contractor, of course, saw it quite differently and maintained that the client should provide both the flanges and the coating. The intent of the contract was quite clear to the engineers but in the end, the contractor won the day. Only about $40,000 was involved, but the interesting point is that the client gave in, not because the contractor was right (in fact, the engineers on both sides knew very well what the intent was) but because the contract language was inadequate and the contractor was persistent.

Analysis:

This claim should have been rejected by the client.

- The contract should have listed material to be supplied as free issue which would not have included the flange coating material.
- The scope of work should have obliged the contractor to supply all installed and consumable material unless otherwise expressly mentioned in the contract.
- The contracts administrator, when gathering data for the bid package would no doubt have passed the documents to his colleagues for double checks and proofreading.

Even if none of these measures took place, the client could reject the claim on the grounds that the plain intent was to have the contractor coat the free issue anchor flanges. Furthermore, all bidders had the opportunity to question the wording in the bid package at the job explanation meeting or at the bid clarification meeting. However, it seems that the contractor triumphed because he pursued the claim and because the contract language could have been misleading. It must be remembered that this is a small claim; too small to be the subject of arbitration or law, where it would have probably failed. In the described circumstances, as an irritation factor, the claim scored out of sheer client exasperation.

Moral: If the client does not take the time and trouble to write the contract as succinctly and as thoughtfully as possible, the contractor cannot be blamed if he takes advantage of it.

CLAIM No. 2. CATERING CHARGES

Caused By: Poor Proofreading

In this contract, the omission of a word, probably a typing error that was not corrected during proofreading, also cost the client a little time and money. It was a lump sum contract in a remote area and the contractor was expected to erect barrack type accommodation for his personnel and also to feed them. The client decided to write into the contract provision for a number of his project staff to be housed and fed by the contractor within the contract price. The scope of work read as follows:

To Be Provided by the Contractor:

All accommodation, messing, catering, and utilities required for Contractor's operations. Provision of such accommodation shall be to Contractor's account.

Contractor shall provide reasonable and adequate messing and accommodation for up to twelve Company supervisory personnel (single rooms) to the same standards as provided to Contractor's senior staff . . .

The first thing to be noticed here is that the last paragraph does not actually state that Company accommodation etc. shall be to Contractor's account but the contractor did not argue this point. He instead presented a bill for client personnel food, on the grounds that the word "catering" was missing from the second paragraph.

"Messing," he declared, "meant the provision of eating facilities only, otherwise why would the scope mention 'catering' in the first paragraph?"

The client's first reaction was to reject the claim, pointing out that his project team would not have intended to bring their own food in this remote area, indeed they could not have done so. The intent of the contract was clearly to have the contractor accommodate and feed them. Possibly there had been a typographical error in the preparation of the contractor but the intent was quite plain. The contractor replied that he was quite prepared to feed the client's staff and in any event, the mess hall was already in existence and used by his own personnel and the cost of twelve more places was included in the front end mobilization costs. The food had to be an extra cost and the contractor had read the contract in this manner, assuming that the word catering was deliberately left out of the second paragraph for this purpose. The matter dragged on for several months with the contractor presenting invoices for all the client staff meals. Eventually, a compromise was reached and the contractor received a percentage of his claim.

Analysis:

Obviously a typo that slipped through the net. The contractor should not have received any payment, but his dogged determination paid off.

CLAIM No. 3. THE GREAT ASPHALT ROBBERY

Caused By: Utter Confusion in the Choice of Contract Format

It has been observed that no oil industry construction contract of any substance is entirely claim proof, but now and again one comes to light as ridiculously vulnerable. In these gems, the contractor does not even have to claim—he just sits back and lets the money come to him!

A Middle Eastern contract was let for site preparation and paving. The contract had the normal Articles of Agreement, followed by a brief scope of work involving earthmoving about one million cubic yards, bringing to grade, and paving with asphalt. The compensation section gave dayrates for equipment and operators. There was no schedule in the contract. The work went painfully slowly. The contractor's invoices came in every month, based not on progress but on daily hire. Somehow final grade was reached, several weeks behind the client's schedule. When the first part of the paving was tackled, it was below specification (also not in the contract) and although the contractor had the right equipment, it became obvious that he was not very expert with it. The client's civil superintendent had the contractor do the paving over again. When the invoice came including the second attempt, the client at first rejected it. The contractor objected. "You have my equipment on daily hire," he said, "it is the responsibility of your supervision to see that the work is up to specification. If it has to be done again, you should pay for it." The client took another look at the contract. It seemed that there was no obligation on the part of the contractor to do anything but to supply and operate his equipment and the contract was, in effect, an equipment rental agreement. "Alright," said the client, "we will terminate your contract under the relevant clause in the Articles."

"You can't do that," replied the contractor, "I have a scope of work and I am entitled to be given the opportunity to finish it and since there is no specification in the contract, I cannot be in default for being behind schedule or out of specification." In the end, the client had to buy the contractor out of his contract. What had happened, of course, is that the contracts engineer who put the contract together started out with the intention of writing a normal equipment rental agreement, but got sidetracked into the inclusion of a brief scope of work. This altered everything.

Analysis:

The first thing to be observed is that the work took place in a developing country where the client engaged indigenous contractors who possibly had the benefit of preferential treatment in these matters. However, this does not excuse the contracts engineer who drew up the agreement documents and seems to have made a complete mess of it.

CLAIM No. 4. THE REIMBURSABLE COST OFFSHORE BONANZA

Caused By: 1. Insufficient Preconstruction Site Investigation
2. No Pre-award Inspection of Contractor's Equipment
3. Lack of Client Control Over Cost and Schedule
4. Poor Organization of Contractors' Work Program

It is small wonder that clients try to avoid agreements involving cost reimbursable contracts. This type of transaction can get wildly out of control even with a cooperative and not necessarily avaricious contractor doing the work. The whole object of the cost plus option is to have the contractor perform for a fee, fixed or otherwise, when the exact nature of the work is not fully disclosed and design and engineering completed is at a minimum. Some cost reimbursable contracts are satisfactorily administered with strict financial control and frequent audits, but now and then an oil company will embark upon a venture which develops into a nightmare of terrifying proportion. Such a situation was encountered when an oil producer needed to extend gas lines from an onshore terminal to a sea island six miles offshore. Several methods were considered. The accepted design involved a concrete trestle on driven concrete piles.

The trestle was composed of 1500 precast, post tensioned concrete cylinder piles, grouped in 250 bents, with an average bent-to-bent span of 125 ft. The piles supported 300 precast pile caps weighing up to 240 tons each. The pile caps in turn supported 900 precast concrete girders as deck sections weighing up to 230 tons each.

A cost reimbursable contract was awarded to one contractor for the offshore concrete construction of the piles and trestle and a lump sum contract was awarded to a second contractor for laying the gas lines on the trestle. Two factors were apparent at the onset. One was that the progress of the trestle construction would be controlled by the pile driving rate and the other was that the schedule of the pipe laying contractor would be governed by the progress of the trestle contractor. The latter arrived on site with his fleet consisting of a jack-up pile driving barge, a heavy lift barge, and various other tugs and barges. The estimated cost of the trestle contract was $40 million. Pile driving into the seabed proved more difficult than anticipated and it was decided to predrill the pile locations. A drill barge was engaged for this purpose together with more floating equipment. There were still severe problems with pile installation but in addition, the predrilled holes began to collapse and fill up with sand. Drilling mud was selected to prevent this, which necessitated more expensive equipment.

A test of grout pile installation procedure was conducted at one location and during installation, the tug serving the grout barge ran aground and the barge drifted towards the shore. The pile barge set its pile hammer on the pile, waiting in this position for the grout to set. When the pile was released after two days, it fell over and onto the pile driving barge. Every offshore contractor can confirm that troubles, once they appear, come not singly but in battalions. So, the heavy lift barge went to the assistance of the pile driver and in the process damaged its crane boom. A supply barge went adrift and collided with one of the more successful pile driving efforts, breaking off the pile below the mud line. A tug towing a work barge needed to get to the other side of the trestle, and rather than going all the way around the outboard piles, decided to steam under the trestle as a shortcut. The tug made it, but the work barge did not; it collided with the piles, knocking them over and bringing down the trestle beams. It just happened that this was the first section to be given to the pipe laying contractor so as not to hold him up unduly.

When the trestle was only 50% complete, its estimated total cost was $140 million, three and a half times as much as the original forecast. Nothing seemed likely to reduce this figure and so the work lurched on to some sort of conclusion.

Analysis:

Basking in the wisdom of hindsight, it can be safely suggested that most of the extra cost of this fiasco and the pipe laying contractor's inevitable claim could have been avoided. Precautions could have included a thorough seabed investigation before construction began, a better inspection and selection of marine equipment before contract award, the preparation of approved work procedures before startup, and a more efficient organization of the pipelay contract so that the contractor was not allowed access until a clear run was guaranteed, and so on, and so on . . .

There were no incentives to control costs to maintain schedule; in fact, incentive seemed to be provided to do exactly the opposite as the further the schedule slipped, the more costs increased. The painful progress of the trestle contractor through his learning curve was paid for by the client, although the contractor profited through rental of his equipment and employment of his personnel in addition to his fee.

CLAIM No. 5. THE CHARPY AFFAIR

Caused By: Failure to bring the attention of the bidders important information vital to the satisfactory performance of the work

This case history describes a successful prosection of a claim by a contractor for failure by the client to disclose information necessary for the satisfactory performance of the work, although it would seem that the contractor had no case at all. It is not known what the outcome would have been if the matter had been referred to arbitration or the law. The client settled before this stage was reached and the case illustrates that some oil companies often prefer to pay a claim quietly rather than parade their embarrassment publicly.

A pipeline was being laid in the cold land of the trolls, over hills and through valleys and under fjords. The people of that chilly country had very little experience with major pipeline construction, so they invited foreign contractors to bid for the job. The design engineering was carried out by a world renowned engineering company who produced voluminous specifications for the work. Each circumferential weld on the line was to be subjected to 100% radiography (nondestructive testing) during construction and the terms of the contract allowed the client to order cutouts of one weld in every thousand for subjection to destructive testing. One of the destructive tests specified was the Charpy V notch impact test.

In pipeline construction, as in any other high quality welding operation, the contractor must submit for the client's approval the detailed welding procedure he intends to employ to join one piece of pipe to another. The proposed procedure must pass visual and nondestructive test (X-ray) inspections and destructive tests in order to be qualified. All this must be done before construction commences and since the contractor would probably be in the mobilization stage at that time, the search for a suitable procedure would most likely be carried out in controlled conditions away from the job site. However, the contractor would be expected to ensure that the procedure would also work in actual field conditions.

Having passed the visual and nondestructive inspection, the test weld is subjected to various mechanical tests. In this case, these consisted of a test for tensile strength, a hardness test, and a Charpy V notch impact test.

In preparation for the latter, three samples or "coupons" are cut out from the test weld. From each of these samples, a small specimen of predetermined dimension is taken and a notch is cut in the specimen. One end is held in a clamp while the other end above the notch is struck with a weight on a swinging pendulum until the specimen fractures. The energy required to fracture the specimen is measured in joules (one joule is equal to 0.7375 foot/pounds). The impact energy required for the pipeline contractor's test weld was an average of 45 joules, or just over 33 foot/pounds. Three tests were taken on the specimens, one at the center of the weld, one at the fusion line between the weld metal and the pipe, and the third at the root weld, which is the first pass made by the welder between the bevelled pipe ends. The first procedure failed the test at the root, but after several attempts, a procedure was found that satisfied all three tests. This procedure was duly qualified and construction began. When the client exercised his right to order a cutout of a field production weld for the purpose of destructive testing, all laboratory tests were acceptable with the exception of the root Charpy results. All production welding was halted until this problem could be solved.

The contractor claimed for direct and consequential costs on the following grounds:

1. Although the specification regarding the Charpy test was included in the bid package, no attempt was made by the client to emphasize the difficulty of meeting it.
2. The Charpy test on the root weld was an unusual requirement even for cold climates, and was not called for in any other part of the world, including Alaska.
3. The root weld will always be vulnerable to impact tests. It is the first run laid and therefore without the benefit of previously deposited hot metal. The root weld is not the strongest part of the finished fused metal and does not, therefore, warrant the severest test.

The contractor maintained that the client had failed to "disclose information necessary to satisfactory performance" by not giving more prominence to the Charpy requirement during the bid stage.

The claim might have died a natural death had the client rejected it, declaring that the contents of the relevant specification were in plain view throughout the bid stage and not hidden in the small print. In any event, a contractor of international repute should have seen that the welding procedures would need special attention and unusual or not, that was the requirement. If necessary, sufficient compensation for the extra work involved should have been included in the contractor's bid. But the client said none of these things and instead started a chain of events that proved very expensive indeed.

To be truthful, the client's team did not emphasize the importance of the Charpy test at the root at the time of bid, or stress the difficulty in achieving it because they did not fully realize it themselves. Since none of the original five bidders raised the subject either, it may be assumed that everyone missed it.

In an effort to get production moving again, the client went back to the design engineering contractor who wrote the specification to see if it was really necessary. Since the DEC was not too insistent on its retention, they decided to issue a waiver to the contractor and reduced the average joule value to 27. The welding procedure now easily passed the test and production resumed. The contractor entered a modest claim of about $800,000 for delay from the time production was halted to the time the waiver was given. Liability was not disputed by the client.

Disaster struck from an unexpected quarter, however, when a government agency nominally responsible for all land transportation of combustible substances decided not to agree with the waiver. Whether this came about as a result of bruised feelings by not being consulted or just bureaucratic cussedness is not clear. Production was stopped again while a new search was made for a procedure to suit the original specification. Meanwhile, back on the line, welders were chafing at the enforced inactivity and loss of production bonuses; a few quit, seeking work elsewhere. A mood of despondency settled over everyone and production in civil work and other areas suffered as well. Winter approached and the pipeline route was covered in a blanket of snow.

After experimenting with several welding procedures, one was found which successfully passed the laboratory tests on the root weld and production slowly started up again. But this time a massive claim was launched by the contractor and the client ended up by paying almost double the original contract price. The hawks in the client's organization recommended payment only from the time the waiver was given to the time it was withdrawn, for the following reasons.

- The requirements of the specification were known to the contractor at the time of the bid.
- No objection to this requirement was raised before contract award.
- The contractor is responsible for the efficacy of the welding procedures in field conditions as well as in shop conditions.
- To assist the contractor, the specifications were revised but following the objections of the government agency, the client was obliged to revert to the original specification. When the contractor resumed production based on the original specification, he was,

in effect, back to square one and it became his responsibility to find a welding procedure that would satisfy the original requirements. Indeed, had the client not given the waiver, this would always have been the case and the delay in finding the right procedure would have been entirely at his expense.

In the end, however, the doves prevailed and the claim went higher and higher up the management pyramid, getting more and more sympathy on the way up, until the contractor made out very nicely indeed!

Analysis:

An interesting contractual question arises from the Charpy affair and that is: Once a waiver is given on a technical specification, can it be withdrawn? In other words, if a test mentioned in the specification is considered to be unnecessary, then technically it must remain unnecessary. Although the client has every right to reintroduce the test, he should be doing so at extra cost. Whatever the answer to that one is, the fact remains that granting a waiver to ease the contractor's burden is a step that should be taken only after considerable thought and after checking all the possible traps. Viewed in hindsight, one may wonder why the client did not check with the government authority before issuing the waiver, but everyone knows the value of hindsight.

The overriding objective in this project was to get the pipeline finished or on schedule before the onslaught of winter. To realize this goal, it would seem that a certain amount of compromise would be in order. On the one hand, the contract documents clearly describe the weld procedure test requirements. On the other, the chain of events showed that those requirements were crucial to the success of the project and deserved to be highlighted during the job explanation and bid clarification meetings. This was not done.

If the test on the root weld did not need to be so stringent, why did the DEC include it in the first place?

CLAIM No. 6. TRAINING LOCAL WELDERS

Caused By: Weak Presentation by Contractor

In preparing a change proposal that has great potential for a claim, contractors should be able to refer to well kept daily logs so that the facts are presented correctly. This is of vital importance. If one wrong fact is discovered by the client, the whole case may collapse.

This happened to a well-known contractor who had hired a consultant to present a claim of some $250,000. This was compensation for having to hire local welders to work alongside expatriate welders, and the subsequent high repair rate suffered as a result. It started off as a fairly legitimate proposal. The contractor had been awarded a job in the Middle East involving a considerable amount of line pipe welding by imported welders. At the time of bid, no mention had been made regarding the necessity of employing indigenous welders. Halfway through the contract, the client, under pressure from the Department of Labor, insisted that local welders be employed and if necessary, given training to improve their skills and pass the welding qualification tests.

The contractor's consultant produced a well argued case that training had cost money that was not in the bid. The client conceded that this was so and offered to examine the matter sympathetically. Encouraged by this reaction, the consultant tried to pile it on a bit more by suggesting that, even when trained, the local welders caused more work for the repair crews than the expatriates. This proposition was also received by the client without much objection but the consultant was asked to supply backup figures. The consultant responded: "In every 1000 welds, 500 by expatriates and 500 by locals, we can show that the repair rate for the former is 10% and 15% for the local welders. Therefore, the locals create 5% more repairs than the expatriates. We can also demonstrate that a repair crew and its equipment costs us $4,000 per day and they accomplish 8 repairs a day, so each repair costs $500. In every thousand welds, an expatriate would expect to cause 100 repairs and the local welder 150 repairs which is 50 more than the imported welder. If each repair costs $500, then this part of the total claim is $25,000 per 1000 welds."

"Not so," said the client's team, "the locals did not weld 1000 joints in your example, but 500, so the rate of repair is 50 for the expatriates and 75 for locals, which is 25 more than the imported welders, equalling a claim of $12,500 per 1000 welds." The contractor's consultant began to argue the point, but without much conviction, and eventually had to agree with the client's version. This started a closer examination of the rest of the claim, as a result of which several more flaws were discovered and the whole case began to crumble.

Analysis:

In the end, this contractor was fortunate to get away with about half of his original proposal. If the consultant had admitted the mistake right away, the client's team would have felt good about discovering the error and would probably have let the rest go unchallenged. This is not to detract from the value of having a consultant or claims expert handle a contractor's claim. On the contrary, the return from this specialist assistance should more than pay for the cost of bringing the claim. The introduction of this sort of help onto the job in the early stages of the contract is highly recommended to supplement income by well aimed change proposals. Contractors should also remember that, in most cases, the client's site team is not the ultimate judge. If the contractor perseveres, the client's team will have to refer the claim to higher management. On balance, the claim may stand more of a chance then, because management knows how much money is in the budget for the entire project, and they do not know as much as the site team about the weaknesses in the contractor's arguments and are more likely to be sympathetic.

CLAIM No. 7. THE "LIFE" OF THE CONTRACT

Caused By: Disagreement Over Hourly Labor Rates

In most of the claim histories related in this chapter, it seems that the client never wins and inevitably pays out. However, the battle is not always so one-sided. Here is a dispute that went to the very brink of arbitration. It was a lump sum petrochemical construction contract and in accordance with normal practice, the bid package had included provision for hourly labor rates for changes and extra work. The unit rates against each discipline were, of course, supplied by the contractor in his successful bid. It is worth reproducing the page of the contract that deals with these rates for potential extra work and authorized contract variations.

HOURLY LABOR RATES FOR CHANGES AND EXTRA WORK

Contractor submits herein its all-inclusive hourly rates for all classifications of labor up to and including supervisor. These rates shall be used to calculate the Day Works Labor Rates for changes or extra work where existing lump sum or unit prices do not apply.

Supervisor mechanical	$55
Supervisor piping	$50

Foreman Mechanical	$45
Foreman Piping	$45
Foreman Rigging	$45
Foreman Scaffolding	$45
Welder	$45
Fitter	$40
Millwright	$40
Scaffolder	$35
Laborer	$30

Overtime—	
Monday thru Friday	Standard Time Rate + 23%
Saturday	Standard Time Rate + 33%
Sunday	Standard Time Rate + 45%
Standby Rate	95% of Standard Time

The Labor Rates given above include for the following:

(a) Labor—wages, piecework, other allowances (travelling, wet money, etc.), all statutory and other related charges such as National Insurance, Graduated Pensions, Insurances, Company Pensions, Holiday Pay, Sick Pay, Workmens' Compensation and Employer's Liability Insurances, Industrial Training Levy, Redundancy Payments Contributions, and any other similar charges.

(b) Small tools such as ladders, chisels, handsaws, buckets, hammers, hard hats, protective clothing, and other items of a like nature.

(c) Consumables—gases, welding rods, oils, lubricants, electricity, and other items of a like nature.

(d) All maintenance charges on facilities, plant, and equipment used and required.

(e) All supervision and management necessary.

(f) All establishment charges, overheads, and profit.

(g) All necessary engineering services.

(h) All temporary facilities, personnel, transport, and the like.

(i) All delay and disruption caused as a result of performing work on an hourly rates basis.

This type of format is in general use throughout the petrochemical construction industry and it is recognized that the rates include for all burdens, overheads, management charges and profit. Contractors are usually quite happy to enter into hourly rate conditions since they are careful, at the bid stage, to negotiate rates which will bring them in a reasonable return over and above the lump sum consideration. Indeed, most contractors would be delighted to continue on an hourly rate basis for the remainder of the contract.

In the dispute we are presently examining, the project had fallen badly behind schedule, mainly due to late arrival of company supplied material. The contractor suggested that the lump sum arrangement should be discontinued and the work should proceed on a time and material basis using the quoted hourly rates, but he advised that the rates would no longer include for overheads and profit. When asked to justify this stance, the contractor explained that as the last milestone date in the schedule had been reached and passed without realization of the intended scope of work, through no fault of his, the unit rates as quoted in the bid were no longer valid. The cost of overheads and profit were contained in the lump sum and were estimated to take effect only over the "life of the contract." Since the last proposed milestone date had been passed, the amounts allowed in the estimate for burdens and profit had been consumed and although the contractor would agree to continue work on a time and material basis using the existing unit rates there would be no provision in the rates for staff above the field supervision such as engineers and management and the cost of supporting this establishment.

The client's view was that there is no contractual term that recognizes "the life of the contract." According to the terms of the contract, the client had the right to order the contractor to continue with the work as long as there was work to be done and as necessary to complete the job in hand. The very purpose of the hourly labor rates was to accommodate extra work, and extra work may mean running over the schedule term of the project. There was no provision in the contract for altering the unit rates when the date of the last proposed milestone had been reached. There was also no escalation clause in the contract which would permit the contractor to apply for an increase or a revision of rates after an agreed period of time. The contractor, upon signing the contract, had agreed to carry the hourly labor rates through to the satisfactory completion of the work. The client believed that the contractor was well aware of this fact at the time of bid. It would be beneficial if contracts would be completed on or before schedule but unfortunately, this does not happen very often in the real world.

Contractors know this; they also know that extra work generates extra profit and they develop their unit rates accordingly. During bid evaluation, the client examined the bidders' proposed unit rates very carefully, as it was anticipated that there would be extra work. There was no indication that the contractor's proposed rates were not commercially sound, indeed, they were on a par or even higher than those of the other bidders.

The client further pointed out that the Articles of Agreement in the contract give the client the right at any time to order changes in contractor's work including increases and decreases. There is no provision that indicates that time is of the essence; therefore, the contractor cannot abandon the work or impose changed rates for time reasons. The governing factor is that the contractor is still doing work and has not demobilized. The contractor is not entitled to say that the contract is at an end. The contractor has a clear contractual obligation to complete the work.

Both sides seemed to have reached stalemate at this point. Further meetings were apparently of no value at even the highest level and although the contractor continued working, albeit slowly, preparation was made to bring the matter before an arbitration panel.

Analysis:

The suggestion to convert the lump sum condition of the contract to a time and material arrangement came from the contractor, so obviously the conversion was acceptable to him provided the hourly labor rates no longer included overhead and profit. This meant that the cost of management, engineers, site facilities, transport, and profit would be extra to the rates. The contractor's case was based on the premise that the rates were originally calculated to include overhead and profit up to the scheduled completion date of the lump sum contract. Presumably, if this date had not been realized due to default by the contractor, then he may have been obliged to continue on the hourly rates as written in the contract. But the delay was occasioned mostly by the failure of the client to deliver free issue material on time. The life of the contract had expired according to the contractor and a new arrangement was in order.

However, unless the contract specifically states otherwise, the scope of work must be completed and in the absence of an escalation clause, work must continue at the contract rates. An exception to this would be if the client ordered work to be carried out that was completely unrelated to the scope: for example, the client proposes building a facility outside the perimeter fence for a purpose not connected to the plant process; a dock or a heliport perhaps. This case history has deliberately paused at the brink of arbitration to enable the reader to make conjectures as to the likely outcome of the hearing.

CLAIM No. 8. THE SMALL PRINT

Caused By: Contract Signed in Haste and Repented at Leisure

In this case, two conditions in the contract documents were completely overlooked by the contractor in his haste to execute the contract. During the performance of the work, the contractor was reluctant to sign the client's standard form change orders because the wording included the following: "The price and time extension granted under this Change Order constitute payment in full for the work described above, including without limitation, all direct and indirect costs and profit and all effects of this Change Order direct, indirect, and consequential on all Contract work whether or not changed by this Order." The contractor, although carrying out the work, could not see how he could sign the above proviso and still safeguard his rights to claim for future costs and impacts not foreseen at the time. Another dispute occurred when the contractor's invoices came in for work done and they were not met by the client within the statutory thirty days. This caused the contractor to complain bitterly and threaten the imposition of interest charges on the unpaid amounts. The client, however, pointed to the relevant clause in the Articles which stated that "payment of contractor's invoices will be made thirty days after approval by the client." The period of approval was not specified. Another example of the contractor not reading the contract before execution?

Analysis:

Since samples of the change order and amendment formats were included in the bid package, the contractor may have negotiated a codicil to the change order which would have enabled him to reserve a provision for indirect costs and future unforeseen cumulative effects on the work. He should also have noticed the wording regarding approval of invoices and realized that this gave the client an unfair advantage in the legitimate delay in settlement of invoices.

The truth is that contractors are in such an unholy rush to get the job that they just do not study the contract language. It is not that the client wishes to trap them; indeed, he is probably not aware of the existence of the pitfall in the bid stage and would willingly alter the wording at that time. It is only when the honeymoon is over and the client sees a way to bring a recalcitrant contractor to heel that he grasps at the opportunity.

Moral: When in doubt, read the contract!

CLAIM No. 9. ENGINEERING SERVICES CONTRACT

Caused By:
1. No control of time sheets and backup data
2. Failure to deal with change proposals in a timely fashion
3. Delay in technical evaluation of vendor bids
4. Unnecessary delay in approval of documents
5. Delay, disruption, and extension due to inefficient working created by circumstances 1–4

Some contracts for engineering services are conducted on the basis of cost reimbursable hourly unit rates for various disciplines such as management, process engineering, civil, plant design, instrument design, electrical design, procurement, and possibly quality assurance and inspection. All these rates should include for salaries, overhead and fees, and all management burdens. However, the problems arise not from the negotiated rates but the man-hours against which these rates are applied.

Of all the claims presented to the client, none is more difficult to combat than the engineering services claim for man-hours expended on design and engineering, particularly when the work is done in the contractor's home office, far away from the construction site. Who can say with certainty that a drawing of a small substation should take ten hours to produce? The draughtsman may admit that, without interruption, he could do the job in half the time but on this particular occasion, he had to wait for certain information from the client. The procurement manager, when told that his invoice for purchasing an item of equipment is excessive, may point out that the client's decision on the purchase arrived too late and it was necessary to accept certain penalties as well as placing his department in an overtime situation. With the construction contract, although the problems are many and varied, in most cases the work may be measured and time sheets more closely monitored. In the field, when extra work is called for as a result of a design change, the contractor will usually be in a position to quote a lump sum. If this is not possible, he will enter into negotiation to give the client a close estimate, preferably before the work starts. When the work is done and disputes still linger, it can be measured or recalculated, and in any case, it is there, in full sight, subject to examination and inspection. Not so with the mysterious ways of engineering services.

One would imagine that if an engineering contractor scheduled a project to be engineered in twelve months from award, the unscheduled client-imposed design changes and delays could be estimated fairly closely and added to the schedule for each contractor discipline. Engineering contractors, however, do not find this completely acceptable and are given to muttering about the domino effect and delaying constraints. The first step in the fight against invoice loading is to concentrate on the man-hours expended and to insist that the contractor produces, at the bid stage, a chart showing the intended manpower loading, discipline by discipline, with peak and trough man-hours. Thereafter, estimated man-hours will be indicated on each design change variation order. The impact on the schedule must also be stated, if there is one. The contractor must be obliged to take into consideration the amount of man-hours expended on each variation in the preparation of his Critical Path Analysis charts. Without this, he is unable to introduce factors for inefficient and unproductive working. Without consideration of the number of men employed in extra work activities, the schedule impact estimate can change dramatically. For example, if the scheduled portion of an activity is estimated to use 80 hours (two weeks) and the extra work bar is shown as 160 hours, this could be represented as an overrun of 80 hours or as having no impact on the schedule, depending on how many men are employed.

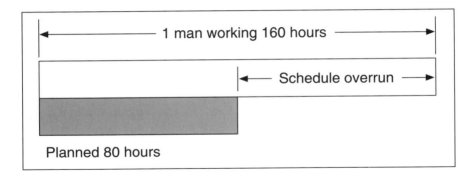

In this diagram, one man is employed on the extra work portion and the schedule is impacted, but if an extra man was introduced, the diagram would not show a schedule impact:

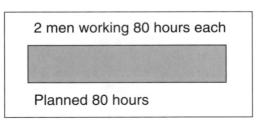

2 men working 80 hours each

Planned 80 hours

As far as the contractor is concerned, the cost of the schedule over-run is not the expense of one draughtsman working an extra eighty hours but includes the whole burden of management at the highest unit rates plus the domino effect in every other discipline.

The engineering services contract, by its very nature, is based on reimbursable cost but closer control may be accomplished by dividing the compensation part of the contract into three sections: reimbursable costs, fixed rates, and fee.

Reimbursable Costs

These are actual net invoiced costs in direct relation to the performance of the contract for consultants, subcontracts, equipment and supplies, and all other direct and reasonable expenses approved by the client to which firm estimates could not be apportioned before contract award.

Fixed Rates

These are fixed unit prices such as the actual cost of wages and salaries paid to the contractor's employees, including social security contributions, unemployment and payroll taxes, sick leave, holidays, vacation, and pensions, but excluding bonuses and profit sharing plans. Added to the fixed rates would be any other costs to which a firm price could be predetermined. Fixed rates are not profit generators.

Fee

The fee would be inclusive of *all* profit, overhead and office expenses, and management burdens and all nonreimbursable costs and profit there on. Nonreimbursable costs include the cost of salaries and travel expenses of executive officers not directly assigned to the project, clerical and secretarial personnel, interest on capital or on borrowed money, income tax, employee bonus, profit sharing plans, and severance costs.

In this case, a client planned the construction of a petrochemical plant. A reimbursable cost contract was awarded to an engineering contractor for the design and engineering of the project. Apart from routine visits to the proposed site, the work was to take place at the contractor's head office in another country.

The contractor complained of disruptions in progress brought about by lack of information from the client, changes in specifications, and delay in decisions relating to procurement of equipment for the plant. During the course of the contract, a number of change proposals were submitted by the contractor and these were disputed or rejected outright by the client, put aside and ignored. The contractor moved into a claim situation. The client finally then agreed to examine each proposal but reluctantly issued change orders for only two or three out of thirty change proposals. Following a number of negotiating sessions, some of the claims were met or negotiated but most remained in dispute.

Undoubtedly, the contractor was asking too much compensation for the client's procrastinations and may even have been guilty of charging for time not spent, but the client had only himself to blame for losing control of the contractor's performance.

Analysis:

From the beginning of the project, the client should have introduced audits and controls, including properly authorized timesheets for all the contractor's employees. Approvals for procurement from vendors and subcontractors should have been made without delay. Change proposals should have been dealt with immediately as they were presented. The client should have insisted on the monthly update of schedules and budget reports.

CLAIM No. 10. PIPE FABRICATION AND THREATENED WALKOUT

Caused By: Attempted Shortcut in Contract Preparation

We now examine a claim presented to a petrochemical plant client by a contractor awarded the mechanical portion of the project. This construction contract was of a pattern common to the petrochemical and allied industries. It consisted of five parts:

The Form of Agreement
Exhibit A–Special conditions
Exhibit B–Scope of work
Exhibit C–Schedule of prices and payment
Exhibit D–Technical specifications and drawings

The contractor involved was experienced in large and international contracts and was familiar with the general composition of this type of agreement. He received a request for quotation along with five other bidders. During the bid clarification meetings, the bidders raised only minor queries and disagreements with the bid package and these were quickly resolved. The scope of work in the contract called for a large quantity of piping to be installed in the construction of the new plant and the contractor was to be responsible for the production of isometric drawings, procurement of material, installation, painting, and testing. The unit rates in the contract Exhibit C provided for the work per linear meter. The pricing depended on material classification, nominal bore size, and the complexity of the area in which the piping was to be installed. No major problems were encountered until it became necessary to make a number of design changes.

Changes to isometrics and the subsequent installation procedures were adequately covered by separate unit rates where the change involved an increase or decrease in piping lengths. The client's home office contract engineer who compiled the bid package decided to simplify the change order procedure by insertion of the following paragraph:

"If a change neither involves an increase nor decrease in the length of pipe to be installed by CONTRACTOR nor effects pipe that has already been fabricated then CONTRACTOR agrees and accepts that no adjustment will be made to the total compensation to the CONTRACTOR; CONTRACTOR having included allowances for such changes in the CONTRACT PRICE."

In other words, if a design change occurred before the contractor had started to fabricate the spool, the contractor would agree to reissue the isometric, fabricate, and install at no extra expense other than the contract unit rate provided the length of pipe had not increased, and by the same reasoning, would not compensate the client if the change decreased the length of the pipe. No objection was raised to this proposition at the preaward meetings.

Unfortunately, the contractor did not anticipate situations where the line diameter would be increased or the pipe class upgraded. According to the terms of the contract, a 25mm diameter line in carbon steel could be changed to a 250mm diameter line in stainless steel and if the pipe length had not been increased, there would be no adjustment to the price. In fact, no such extreme changes were made during the project. Several diameter and class changes did occur that did cost the contractor money, but according to the client, they were included in the contract price. The project was already behind schedule and over budget and the client, having previously conceded a number of other and non-related claims to the contractor, was not in the mood to capitulate on this one. Nevertheless, the claim had to be investigated and the cost engineers set about remeasuring the isometrics and checking unit prices and cost of materials. On paper, there was only a small difference between the total amount of the contractor's claim and the client's calculation when the contractor first realized the potential impact of the offending clause. However, the project was not finished and it was accepted that there would be further design changes. Much correspondence had been exchanged on the subject, culminating in the client's last letter of rejection. However, the contractor refused to accept the client's decision and requested a meeting on the claim, hinting darkly that a complete shutdown might occur if the clause was not amended.

The client's team consisted of the contracts manager, the contracts engineer in charge of the contract, and the field piping superintendent. The project manager was to sit in as an observer. The contracts manager was the only spokesman for the client's team and if the others wished to make a point, they did so through him. At a previous and private meeting, the client's team had discussed the subject and decided on the following ground rules:

- The contracts manager is to be the only voice on behalf of the client.
- Basic courtesies are to be observed at all times and a calm atmosphere maintained.
- Contractor is to be invited to speak first
- No minutes are to be taken but a record is to be made of any agreements or decisions reached.

- If any separate discussions are necessary, contractor's team is to leave the conference room for the purpose.
- Meeting venue is to be the client's conference room.
- Client's agreed tactics are to adhere to the contents of the last letter on the subject (i.e., rejection of the claim), at least at this initial meeting.

At their unofficial meeting, the team also discussed the likely outcome if the contractor insisted on some settlement of the claim and they agreed that a work slowdown, or even a complete stoppage was possible. It was decided to seek the advice of the company lawyer.

The legal opinion was:

1. The client was properly entitled to make alterations under the terms of the contract and to give instruction for additional work provided it clearly related to the original scope of work. By so doing, the client had not materially breached the contract in such a way as to entitle the contractor to refuse to carry out the work or leave the site.
2. Even if, having given the work change instructions, the client is unable to agree with the contractor on any extra amount in compensation, the client is entitled to require the contractor to carry out the work.
3. Payment of compensation or otherwise may then be dealt with through arbitration as provided for in the contract.
4. Should the contractor obviously adopt a slowdown policy or leave the site there would be no possibility of obtaining any form of injunction to prevent that. Financial claims against the contractor would necessarily have to exclude all consequential loss but could include all extra direct costs, which would be substantial. These direct costs would include the expense of appointing another contractor to finish the work.
5. Costs of other subcontractors being delayed by a walkout by the contractor would not be consequential damage but would be claimable under law.

Armed with this information, the client prepared for the first meeting.

The contractor's case was as follows: Contractor accepts that the contract excludes compensation for changes that do not involve a change in the line length, but suggests that it was not the *intent* of the contract to have the contractor bear the cost of increase in diameter or a change from a relatively low cost pipe material such as carbon steel to a high cost material such as stainless steel or nickel alloy. Contractor believed that changing diameter or pipe class resulted in a decrease in the length of one diameter or pipe class and an increase in another. The client therefore should receive a credit for the omitted pipe and be charged for the replacement.

The client's contracts manager replied that there had been ample opportunity for the contractor to raise this matter at the bid clarification meetings before contract award. A copy of the contract was in the contractor's hands long before execution. It must be reasonable for the client to assume that provision for extra costs related to the relevant clause in the contract was made by the contractor. Regarding the subject clause, it is concise and crystal clear and contrary to the contractor's interpretation, it was definitely the contract *intent* to encompass all costs in all circumstances related to this clause. The contracts manager added, "We have taken legal advice on this matter and it would appear that there is no possible ambiguity regarding the soundness of the contract clause and we are further advised that any attempt to abandon the work by you to the extent that we are obliged to make other arrangements would lead to a claim against you for substantial direct costs. We do, however, recognize your entitlement to seek the appointment of an arbitration board and in order to assist you, we are prepared to confirm that acceptance by you of the contract terms as interpreted by us does not in any way prejudice your entitlement to seek such an appointment. We would suggest that the appropriate course is for the contractual works to be concluded by you with all possible and proper speed, at the conclusion of which we will be prepared to discuss arbitration arrangements."

Faced with this inflexibility from the client, the contractor backed down and finished the project. There was a small sympathy claim presented at the closeout which was accepted by the client.

Analysis:

We are left to speculate about the possible outcome had the matter gone to arbitration. Would the experts really believe that the client had inserted the offending clause in the bid package with the full understanding that it included costs for changes in diameters and pipe class? Quite obviously, the home office contracts engineer had constructed the clause as a device to speed up compensation calculations and had completely forgotten or missed the effects caused by a change in diameter or pipe class. This is a case of a shortcut becoming a short circuit. However, the contract was already behind schedule and over budget. The project manager was probably under pressure from his superiors not to squander any more time and money on this contract. The subject clause may not have included provision for extra payment for increased diameters or classes but on its own, it was undoubtedly crystal clear and all bidders had professed complete understanding at the time of bid clarification. When a project is going well, on time and within the money allotted, it is easier to be generous if the contract is found to be morally wrong but contractually correct. This is not the case when the project is having difficulties and there is a strong temptation to insist that the contractor abides by his commitments.

CLAIM No. 11. UNPROVEN TECHNOLOGY

Caused By: Attempting Lump Sum Format Without Pre-engineering

An important and expensive lump sum contract was planned for the construction of an automated plant. The design engineering had progressed to the extent that a hard money contract could proceed without difficulty except in one section of the work. There was a requirement for a fleet of unmanned mobile units, rather like intelligent golf carts, which would operate along guidance tracks to perform inspection and maintenance tasks throughout the plant at predestined stops.

Preliminary designs involved rail mounted units but this idea, although the least expensive, was rejected in favor of battery-powered vehicles, moving along a "buried wire" guidance path. On occasion, the units would be required to move away from the track, either under remote control or manually steered, hence the second thoughts on the rail mounted plan. The client's outline design, which required development by the successful bidder, involved guide cables which were to be laid in the concrete floor throughout the plant and which would carry a weak alternating current around which an electromagnetic field would be generated. This would induce alternating voltages in antenna coils centrally mounted on the front of the buggies. The voltages would be converted into a signal indicating the extent of the vehicles' deviation from their desired course. A microprocessor would then compute from the signal the necessary correction of steering angle to make the appropriate adjustments to the steering wheels. The onboard computer would control the stops and tasks of the machine. The client did not wish to depart from the lump sum concept of the rest of the contract, although he could (and perhaps should) have made a separate cost reimbursable agreement for the design and supply of the vehicles. Part of the buried wire guidance system could easily be included in the civil works since the route of the buggies was known and planned in advance. This left only the vehicles themselves for which there was just a rudimentary design.

Bidders were invited to quote for the complete work, civil, mechanical, electrical, and instrumentation. For the vehicles, the scope included but was not limited to design engineering, prototype development, construction, and installation. The successful contractor, in accepting the design and supply of "unproven technology," was also accepting the risk of producing a system within the lump sum price. From the contractor's point of view, he was obliged to provide a method which performed adequately and carried out its intended tasks efficiently but at the least cost to himself. He did not intend that the client should have the very best system procurable at whatever cost to the contractor in designing, developing, and supplying such a system. The number of buggies had not been specified in the contract as the intention was that the client would be buying the overall system and the contractor would supply a system fit for its intended purpose. Obviously, it became in the contractor's interest to minimize both the number of buggies and the refinement or extras being built into them.

In accordance with the terms of the contract, both sides were obliged to seek and arrive at a balanced solution. Balanced, that is, between the very best possible and a suitable and workable system. The actual words in the contract were as follows:

"It is recognized by the parties to this Contract that the process of developing and finalizing the design of the Work will require the closest consultation, cooperation, and coordination between them. It is further recognized that it will be necessary for the parties to develop and agree methods and procedures to enable that process to be carried out and to ensure that the Work represents a balanced solution in terms of capital cost without increasing the Lump Sum price or extending the completion date, provided that such developing and finalizing remains within the Scope of Work of the Contract."

The system was designed, built (by subcontract), and installed with very few technical hitches and to this day the little buggies are running around the track happily carrying out their automated tasks. The contractor's problem was that it was costing him more to install the system than he allowed for in that part of the total lump sum. The design and development work involved in arriving at a suitable electronic guidance method, together with the installation of the manipulative gadgets on the buggies proved far more expensive than he had anticipated and certainly more than a conventional rail mounted system would have done. He also maintained that features were included which were not strictly essential to performance efficiency. In other words, the client had insisted on premier quality when a less expensive model would have suited perfectly well in performance and endurance. There was no balanced solution as originally intended. The contractor maintained that he only agreed to carry out this part of the work for a lump sum on the basis that there was some certainty as to the work and the obligations to be performed.

The client had produced the outline design and although the contractor agreed that he was to be responsible for the subsequent design phases through final design, he maintained that the client's role was to approve each phase and if there were extra requirements, to issue a variation order. Instead, the client had insinuated refinements under the guise of suggestions or safety measures. When the contractor raised objections, the client delayed approval and progress payments until the time taken over submitting and resubmitting designs was encroaching on the critical path of fabrication. In addition, the client had frequently failed to provide information required by the contractor at each stage of the design, interfered in the design process on a continuing basis, and refused to observe the contractual time limits for the return of comments on documents submitted.

The contractor intended that the lump sum work would be based upon proven technology in that the contractor would be able to foresee the consequences as to time and cost, of integrating, adapting, and committing such technology to meet the requirements of the project. However, the client required the contractor to exceed the limits of such proven technology while at the same time reasserting the lump sum price constraints upon the contractor. The contractor argued that the validity of the lump sum price provision is regarded as dependent upon the continued existence of the fundamental bases upon which it was originally thought that the contract would be performed. Once those circumstances change, the condition ceases to be satisfied and the fixed price provisions cannot be relied upon. Furthermore, the contractor held, the risk inherent in a lump sum price contract is balanced by the circumstances in which the parties intended that the contract should be performed. If those circumstances change and the balance is upset, the lump sum price ceases to be applicable.

The contractor entered a claim for the difference between that part of the lump sum set aside for the vehicles and the actual cost plus profit, plus a further sum for delay and disruption not only to the design, development, and installation of the vehicles, but to the whole project via the domino effect.

The client rejected the claim in its entirety and defended his position as follows:

1. Any embellishments that the contractor may have added during the design process were all part of the natural development of the work, even if the client had added a suggestion or two.
2. The client's preliminary outline design was not intended as a tender document but as an indication of the client's approximate requirements. It is not a condition that a lump sum project should be defined to any particular degree of precision at the time of contract formation.
3. Any delay and disruption arising from development of the design was not as a result of any failure by the client but was the direct result of a general failure by the contractor to adequately perform its obligations.

   Regarding the so-called domino effect, it is standard practice on construction projects to assess the changes and delays that occur during design and construction to determine if they have caused or will cause any impact to the overall project. To the extent that the project experienced delay and disruption, it was caused largely by the contractor's poor initial design effort and failure to design a system that satisfied the contract's performance requirements.
4. The client did not wrongfully interfere with the design process but simply insisted on an adequate workable design. That it took more time and money to produce such a design is testament to the inadequacy of the contractor's initial efforts.
5. The contractor's account is misleading in that it suggests that the client "instructed" the contractor to carry out design alternatives. The client did not issue instructions, he merely made a request to which the contractor agreed.

6. The contractor ignores its own shortcomings with respect to any delay in the design process. In many of the instances in which the contractor contends that the client "changed" its requirements, the client was simply rejecting the contractor's inadequate design.

7. The contractor had no intention of fulfilling its obligation to provide a "balanced" design but prepared a design that would result in the lowest capital expenditure that could in any way be argued to comply with the contract's performance requirements.

8. Rather than interfering with the design process, the client had worked hard to assist the contractor, frequently helping it to solve its own problems, suggesting viable technical solutions, and assisting in design coordination.

9. The contractor claims that design solutions involving anything other than standard equipment or standard design involves the use of unproven technology. This is an over-simplistic presentation of the realities of construction projects and mechanical engineering design. Some aspects of a project will inevitably call for a unique solution for which special equipment has to be designed and developed. This does not mean that the materials, components, mechanical principles, or design methods used in developing that equipment are unproven. The buried wire guidance system uses proven principles and only the vehicles needed development, although much of this work is already proven technology. The client was entitled to expect a design developed in accordance with good up-to-date engineering practice.

10. The contractor claimed that any "investigation" of an alternative design should be treated as extra work whereas the client held that it was by way of a *quotation* for extra work, to be made into a change order in the event that the alterative design was adopted. If it was not, there was to be no change order.

11. Finally, the client maintained that the contractor had made a great deal of money on the bulk of the contract and even if he lost some on the vehicle design, development, and supply, he was more than compensated through the remainder of the work.

Analysis:

It would be irrelevant to dwell too long on the folly of entering into a lump sum agreement without the benefit of at least some design engineering. It was done in this case history and judgment may only be made on the facts outlined above. The choice was open to the client to arrange separate design and construction contracts for the vehicles but he chose not to do so, preferring a single design-and-build type, hard money contract. Thus he accepted the possibility of constraints on design and construction. The client offered a good defense against the claim and was doing reasonably well until he mentioned that the contractor had made such a large profit from the civil, mechanical, and electrical part of the contract that he could afford to lose on the vehicles. The client should not have argued in this direction. In a lump sum contract, the contractor's cost is no business of the client. Even if the contractor had made a large profit out of the vehicles as well, he would still be entitled to bona fide variation orders or payment of a genuine claim. If the contractor can observe the performance requirements with a lower level of quality and services in his design, then it follows that he is free to do so without reduction in the contract price. Equally, if a higher level is required to meet the performance criteria, then the contractor must provide it without additional recovery.

The client is correct on the subject of unproven technology. It is difficult to imagine how the contractor can succeed with this line of argument. The only major difference between the original conception of the rail mounted vehicles and the buried wire guidance system is the method of keeping the vehicles on their intended paths. Both of these methods have been used before and are therefore proven technology. The parts that gave the contractor trouble were the manipulative arms and gadgets on the buggies, but these were intended for installation on both the rail mounted and the wire systems. The nub of the argument seems to lie in the contractual agreement to arrive at a balanced solution. The expression is out of place in a formal arrangement. It cannot be precisely defined and expects too much of either party to adhere to it. Obviously, the client wants the best possible equipment for the purpose and the contractor does not want to spend too much time and money in producing anything but an adequate system which will perform satisfactorily.

The contractor acknowledged that he was to be responsible for the design phases from contract award and insisted that the client should have approved each phase as submitted. If the client had wished to add or alter anything, these additions and alterations should have been the subject of variation orders. This attitude seems to ignore the balanced solution philosophy but in any case the contractor did not mention that he had submitted change order proposals that had been rejected. The accusation that the client did not respond to submittals within the time limits imposed by the contract has more substance, and would support the claim for delay and inefficient working.

On balance, it would seem that the client has a stronger case than that of the contractor and the argument that circumstances had changed to the degree that the lump sum price was no longer applicable would not appear to be completely logical. Under the admittedly exceptional terms of the contract with its balanced solutions, surely the client is entitled to comment on the contractor's proffered designs without causing delay and disruption, provided he does so within the contractual time limit.

CHAPTER EIGHT

Payments and Contract Closeout

INVOICING AND VERIFICATION

The client is concerned with getting the work completed as specified, on or ahead of schedule, and on or under budget. Whether or not these aims are realized, the daily transactions of the contract must be supported throughout its life with auditable documents, right up to the final invoice and beyond. Unfortunately, fully executed documentation is not always at hand when it comes to paying the bill.

The contractor wants to leave the job with his reputation intact (if not enhanced), but above all, he wants to be paid regularly and on time. The regularity and timeliness of his payments will depend to a large extent on the systems and controls adopted by the project.

No client Finance Department is going to pay the full value of an invoice on presentation and without verification. There must be an agreed method of checking that the work has been done or the milestone achieved. There must also be an agreed period between presentation and payment (about 30 days usually) for verification and clearance. In addition, the timing of invoice submission should be reasonable. It would be absurd for a contractor to invoice every day or every time a yard of concrete is poured or a length of pipe is laid. Payment may be made against milestones achieved, progress per month, or on measured work. In certain situations, special arrangements may be made to allow front-end payments for mobilization, setting up of camps, and the like. Even without these special arrangements, the contractor will be well advised to cram as much opportunity for front-end payments as he can into the agreement at the time of bid. To eliminate delay in payment, a cutoff date should be agreed for the month's work, assuming that payment is to be made against percentage of progress or measurement or both. If the cutoff date is fixed at, say, the twenty-first of each month, the contractor would prepare a proforma invoice to be presented to the site team for approval.

This is where the contractor's cost engineer or quantity surveyor and the client's contracts engineer are able to work in tandem to expedite verification and help sustain the cash flow. On receipt of the proforma invoice, the contracts engineer will consult with the rest of the site team on the progress made by the contractor and will adjust the proforma accordingly. Eventually (but hopefully before the end of the month), an agreed amount is reached and the contractor may then present the invoice proper for payment. Ideally, both proforma and final invoices should be divided into sections or work packages that represent the segments of the contract and total the original contract price, with change orders being kept separately. Payment will be made according to the terms and conditions of the original contract and subsequent variation orders. The Finance Department or Accounts Payable will expect verification of the contractor's invoice by the site team but will also check the contract and change orders to find authority for payment. In certain projects, this detective work is done by an offshoot of the Finance Department know as "Contract Compliance" which also has other functions of investigation, as the title suggests.

WORK BREAKDOWN STRUCTURE

Many client organizations use a "family tree" type of subdivision of the complete project scope of work to provide the necessary framework for the establishment of project baselines for cost. This is known as cost control or work breakdown structure (WBS).

Each contract in the project is identified as a branch of the WBS tree. A recent trend is to extend the WBS into the compensation section of the contract in the form of a series of work packages relating to the priced scope of work. In a $100 million refinery contract, for example, the work package may appear as follows (payment is to be made against monthly progress):

WORK PACKAGE NO.

1. Site preparation	$18,000,000
2. Fireproofing	$ 2,000,000
3. Civil work	$15,000,000
4. Mechanical work	$20,000,000
5. Insulation	$10,000,000

6. Painting $5,000,000
7. Electrical $15,000,000
8. Instrumentation $15,000,000

Each work package will be broken down into identifiable elements of work:

WORK PACKAGE NO.1—SITE PREPARATION

ACTIVITY DESCRIPTION	COST ($000s)	MAN-HOURS
Earthmoving	6,000	15,000
Trenching	3,500	8,500
Access roads	4,500	11,250
Grading	2,000	5,000
Paving	2,000	5,000
	18,000	45,000

Assuming a lump sum contract and monthly progress payments, agreement must be reached on percentage of achievement of each work-package. Table 1 is an example of a typical invoice summary. It is presented in the form of an accumulative value invoice indicating the total cost of work performed to date, less the total value of all previous payments. Thus it reveals the actual value of all authorized work completed in the period:

Table 1
Invoice No. 011 (in $000s) for November 1993

W.P.	Description	Contract Value	Previously Invoiced	This Invoice	Accumulative Total
1.	Site preparation	18,000	18,000	nil	18,000
2.	Fireproofing	2,000	1,000	500	1,500
3.	Civil work	15,000	12,000	1,000	13,000
4.	Mechanical work	20,000	10,000	2,000	12,000
5.	Insulation	10,000	2,000	1,000	3,000
6.	Painting	5,000	1,000	1,000	2,000
7.	Electrical	15,000	5,000	2,000	7,000
8.	Instrumentation	15,000	3,000	1,000	4,000
		100,000	52,000	8,500	60,500

(Table continued on next page)

Table 1
continued

W.P.	Description	Contract Value	Previously Invoiced	This Invoice	Accumulative Total
Change Orders			20,000	5,000	25,000
			72,000	13,500	85,500
Gross value of work from summaries:					85,500
Less amount previously invoiced	:				72,000
Value of this invoice	:				$13,5000

Table 1 shows that, barring change orders, the original contract value is depleted by 60.5%, leaving a balance of $39,500,000. Site preparation is 100% complete and paid for. The following progress was achieved for this month:

Fireproofing	25%
Civils	6.67%
Mechanicals	10%
Insulation	10%
Painting	20%
Electrical	13.34%
Instrumentation	6.67%

The client's team will also be in a position to advise the Finance Department on the probable cash flow position through the completion of the work and what should be reserved for payments in each future month. How many variation orders will be issued is not so simple to forecast but the contracts engineer can probably offer an educated guess.

MEASUREMENT OF PROGRESS

The verification of invoices relating to contracts with unit rate or bill of quantity arrangements may involve site approved time sheets for man-hours, measurements from drawings or profiles, or even physical checks such as weighing station tally sheets for truck loads or transit mixers. Where the contractor is engaged in a unit rate activity in which precise measurement cannot be guaranteed, as for example, in rock blasting and excavating for a pipe trench, physical measurement must be abandoned for a theoretical calculated amount. Otherwise, the contractor would have no incentive to keep to the specification and if he were paid for the rock and spoil actuallly removed, overblasting would be charged to the client's account. In these circumstances, the contract will be written so that the client will pay only for theoretical quantities calculated in accordance with the standard in the area. The contractor will produce a profile and draw a cross section of the trench which is checked against a cross section produced by the client's surveyor. Both parties will sign against the agreed profile as a base for calculation of theoretical quantities. This way, if the contractor blasts deeper than the agreed theoretical measurement, he does so at his expense.

Measurement of progress from construction drawings, profiles or isometrics is always preferable to physical measurement in the field. The drawing measurement is less costly and provides a record against which future progress may be compared. Where project engineering involves the production of a large number of isometric drawings, it is an advantage to have each isometric on the computer screen as soon as it is issued "Approved for Construction" in order that future design changes or revisions may be measured against the original for addition to or deletion from the contract price via a variation order.

RETENTION

It is customary for clients to pay only ninety or ninety-five percent of each invoice presented and retain the balance against the possibility of defective work or even abandonment of the work by the contractor. As each monthly invoice is submitted, the retained amount will accumulate until job completion is satisfactorily achieved. At that stage, half of the retention money will be repaid, leaving the balance to safeguard the work through the maintenance or warranty period, if the contract demands one.

There are exceptions to the retention rule and in certain types of contracts, consultancies, for example, the client may waive the requirement for withholding a percentage of the invoiced amount. A onetime invoice for work of short duration may also be free of retention. Some clients do not extract retention from invoices relating to change orders or compensation for standby periods. Although the latter concession is understandable it is not clear why payment for extra work should be free of retention. It seems fair to adopt the general rule that payment for anything to be incorporated into the permanent works through the original scope or by change order should be subject to retention.

Clients do not normally pay interest on the money they withhold but a contract worth $100 million with a retention of 10% and a life of 12 months could realize over $600,000 if the contractor were able to invest that money instead of leaving it in the client's hands each month. It is, therefore, a considerable saving to the contractor if he is able to negotiate with the client the acceptance of a bank guarantee in lieu of retention. It is an even greater saving if he can arrange a bank guarantee for each monthly payment as shown by Table 2.

Table 2
Savings Realized From Bank Guarantee vs. Retention

Contract Worth $100m	10% Retention = $10m	Probable Bank Guarantee Charge = 1.5% p.a.	Interest on Retention at 10%
Invoices			
1st– 15m	1.5m	22,500	150,000
2nd– 5m	.5m	6,875	45,834
3rd– 6m	.6m	7,500	50,000
4th– 8m	.8m	9,000	60,000
5th– 10m	1.0m	10,000	66,667
6th– 12m	1.2m	10,500	70,000
7th– 15m	1.5m	11,250	75,000
8th– 10m	1.0m	6,250	41,667
9th– 9m	.9m	4,500	30,000
10th– 5m	.5m	1,875	12,500
11th– 3m	.3m	750	5,000
12th– 2m	.2m	250	1,667
100m	10.0m	91,250	608,335

Table 2 assumes that the contract price is loaded at the front end for mobilization payments and for workoad peaks between the fifth and eighth months. It also assumes that there will be no variation orders; or, if there are, no retention or bank guarantee will be levied on them. An imposition of retention on 10%–15% extra work will, of course, make the bank guarantee that much more desirable. Further savings, for the contractor may be achieved if the client will agree to impose retenion or the bank guarantee requirement only on invoices directly related to the permanent work.

A charge of 1.5% per annum of the amount guaranteed is mentioned as the probable fee charged by the bank but much will depend on the contractor's rating. It has been known for contractors to hawk the contract around several banks and emerge with small parcels of guarantees because one bank would not risk the whole package.

WARRANTY PERIOD

When the contracted work has been completed to the client's satisfaction, the agreement may then enter into a warranty period. During the warranty period, the contractor *warrants* that the works completed will conform to the drawings and specifications and that they will be new and suitable for the purpose for which they were intended and will not be defective. The warranty period will begin at the issuance of the completion certificate given by the client to the contractor and will expire usually a year later. If anything goes wrong with the works during the warranty period, the contractor will be obliged to return to the site and put the matter right at his expense. To encourage the contractor in this responsiblity, the client will keep half of the retention on paid invoices until the end of the warranty period. In the case of the bank guarantee, the client will only release 50% of the guarantee. Some clients, on expiration of the warranty period, will issue a Final Acceptance Certificate and at the same time, give back to the contractor's bank or the contractor, the balance of the guarantee or the remainder of the retention.

CONTRACT CLOSEOUT

When the last claim has been settled and the final invoice paid, the client will ask the contractor to sign a release confirming that no further monies are owed and there are no further claims to be considered. This "Certification and Release" sometimes accompanies payment of the final invoice.

The format of the release document may vary between a simple agreement as shown below and a quite formal affair couched in legal jargon and beginning:

Know all men by these presents: For and in consideration of the payment to Contractor of the below balance by Company for labor, materials, equipment, and services furnished in connection with the performance of Contract No. Contractor XXX does

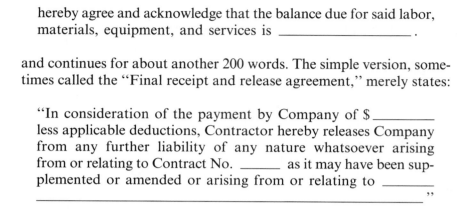

hereby agree and acknowledge that the balance due for said labor, materials, equipment, and services is _____ .

and continues for about another 200 words. The simple version, sometimes called the "Final receipt and release agreement," merely states:

"In consideration of the payment by Company of $_____ less applicable deductions, Contractor hereby releases Company from any further liability of any nature whatsoever arising from or relating to Contract No. _____ as it may have been supplemented or amended or arising from or relating to _____ _____ ,"

Apart from provision for the signatures of both parties that is all there is to it and it seems to give sufficient protection within most organizations.

For the contractor, much of what follows is fairly straightforward and he may silently fold his tents and steal away, laughing all the way to the bank or licking his wounds, as the case may be. For the client, however, there now begins the task of closing out the contract documents, writing the contract history, and recording the problems encountered and the solutions arrived at. Finally, the contracts engineer will complete the "Contractor Performance Evaluation" and when he comes to the question, "Do you recommend using this contractor again?" hopefully he will be able to answer in the affirmative.

The client's contracts administration staff are among the first to be inducted into a new project and the last to leave the field. They generally remain to settle claims long after construction is completed, and of course, they close out each contract. On a large project, the closing out process may take several months. The task of documentation is formidable. This is not to suggest that all the project contracts are saved until the very end before being completely discharged. The site preparation, civil, and other early contracts are closed out as soon as possible during the course of the project. However, the finalization of all documents for the mechanical, electrical, painting, and other contracts in the last stages of the project will continue long after the remainder of the construction team has departed.

For each contract, the contracts engineer will complete a checklist of all actions necessary to close out the contract.

Contract Closeout Checklist

Activity	Status*	Date
1. Contractor's Notice of Completion	_____	_____
2. Outstanding requirements memo	_____	_____
Input received from:		
Project	_____	_____
Construction	_____	_____
• Construction manager	_____	_____
• Site superintendent	_____	_____
Engineering	_____	_____
Procurement	_____	_____
Inspection/quality control	_____	_____
Cost/scheduling	_____	_____
Accounts	_____	_____
Contracts home office	_____	_____
3. All work orders finalized	_____	_____
4. All change orders finalized	_____	_____
5. All amendments finalized	_____	_____
6. Claims resolved	_____	_____
7. Backcharges debited	_____	_____
8. Certification and release	_____	_____
9. Contractor's final invoice	_____	_____
10. Final invoice approvals	_____	_____
11. Final invoice to accounts	_____	_____
12. Final payment to contractor	_____	_____
13. Bank guarantee	_____	_____
14. Retention	_____	_____
15. Contract closeout report	_____	_____
16. Closeout notice to contractor	_____	_____
17. Contractor performance evaluation	_____	_____
18. All closeout documentation filed	_____	_____
19. Final Acceptance	_____	_____

*R - received, I - issued, W - waived, C - completed, T - transmitted,
N/A - not applicable.

On a large contract, this checklist may be in the contract engineer's possession for some time and it will be advisable to prepare a note of activities that must be completed to satisfy each stage of the closeout process. When the contractor believes that he has completed the work that he agreed to perform under the terms of the contract, he will submit a notice of completion to the client. Usually, the notice is required within 10 days of the declared completion.

The contract engineer will liaise with other disciplines to determine whether the contractor has indeed completed his obligations according to the contract. If this is not so, the contract engineer will issue a notice of rejection to the contractor and will not proceed with the checklist until the contractor has rectified the outstanding deficiencies. Following settlement of all claims, backcharges, and other credit and debit balances, all relevant invoices will be reconciled against the contract value together with all work orders, change orders, and amendments. This is done to ensure that final payment is inclusive of all financial arrangements and transactions of the contract.

The contract engineer then prepares the contract closeout checklist which will eventually be distributed and placed in the appropriate file. For his own benefit he will compile the separate notes previously mentioned to assist him in the satisfactory completion of the checklist. These notes may take the form of questions put to himself as follows:

1. Confirmation from site representatives that all tests, inspections, QA/QC matters cleared?
2. All check lists well documented and on file?
3. Any credits to contractor covering the value of work in the checklists that will not be completed by him?
4. Clearance obtained for the following technical documents required by the contract?
   (a) Inspection release certificates
   (b) As built drawings
   (c) As built dossier
   (d) Mechanical completion file
   (e) Other
5. Satisfactory material reconciliation data prepared by contractor with respect to company issued material and contractor furnished reimbursable materials?
6. Scrap materials sold and credited to company?
7. Surplus materials from company issue transported to nominated location, and loaned supplies and equipment returned?
8. No backcharges outstanding?
9. Final value of all work orders, change orders, and amendments agreed with contractor?
10. Retention details confirmed?
11. Bank guarantee details confirmed?
12. All claims settled?

13. Final account prepared and arrangements made for execution by both parties including release of claims?
14. Site files complete and ready for shipment to home office?
15. Site office closed and all office equipment not presently in use packed for shipment or returned to contractor?
16. All delegated signature authorities such as site/company representative cancelled?
17. Contract closeout report and contractor performance report completed? (see Appendix for model performance report)
18. Certification and release issued?

PROVISIONAL NOTICE OF ACCEPTANCE

The client may issue a provisional notice of acceptance in situations where the work is substantially completed and all tests passed to the satisfaction of the company representative. The contractor may provide a written undertaking to complete certain outstanding items by an agreed date. Details of a provisional notice of acceptance, should one be issued, will be included in the contract closeout report which will also contain the contractor's performance report.

CONTRACT CLOSEOUT REPORT

Note: All members of the client's site team will provide information for this report but the contract engineer will be responsible for coordinating its preparation.

Contents

1. Introduction and summary
2. Commercial and finance
3. Contractual
4. Subcontractors
5. Material reconciliation
6. Technical documentation
7. Cost and schedule comparison
8. Contractor's performance report
9. Photographs

1.0 Introduction

1.1 Contractor's name   _____

1.2 Contractor's address   _____

1.3 Contractor's representative   _____

1.4 Description of the work   _____

2.0 Commercial and financial   _____

2.1 Final contract value $_____

2.2 Agreement of final account   Date agreed _____

2.3 Retention

    Release of first half   Date _____

    Amount retained $_____

    Release   Date _____

2.4 Bank guarantee(s)

    Guarantee amount $_____

    Received   Date _____

    Accepted   Date _____

    Return   Date _____

2.5 Accounting – commitment

    Original commitment $_____

    Add/omit to commitment:

      Work orders $_____

      Change orders $_____

      Amendments $_____

      Escalation (if applicable) $_____

    Total revision commitment $_____

2.6 Reason for over/underrun of original

commitment value: _____

_____

_____

_____

2.7 Original contract price $ _____

Total work orders $ _____

Total change orders $ _____

Amendments to contract total price $ _____

Total increase $ _____

Final contract value $ _____

2.8 Insurance claims

Amount claimed $ _____

Amount agreed $ _____

2.9 Construction claims by contractor

Amount claimed $ _____

Amount agreed $ _____

2.10 Counter claims and backcharges by company

Amount claimed $ _____

Amount agreed $ _____

2.11 List of work orders

Date	Description	Amount ($)

2.12 List of change orders

Date	Description	Amount ($)

2.13 List of amendments to contract

Date	Description	Amount ($)

3.0 Contractual

3.1 Contract document

Execution   Date _____

Completion   Date _____

Contract amendments

Description: _____

_____

_____

_____

Execution dates:

001 _____

002 _____

003 _____

3.2 Extension of Time

Initial extension: _____

Subsequent extensions (s) _____

3.3 Milestone completion certificates

Description: _____

Date issued: _____

3.4 Notice of rejection: Date issued _____

Notice of acceptance: Date issued _____

3.5 Certification and release: Date issued _____

4.0   Subcontractors

Scope of work: _____

Name: _____

Scope of work: _____

Name: _____

5.0   Material reconciliation

5.1   Details of scrap materials:

Material: _____

Quantity: _____

Value of sale: _____

Date of sale: _____

5.2   Details of surplus materials

Material: _____

Quantity: _____

Deposition:

Transported: _____

Stored: _____

Location: _____

6.0   Technical documentation

6.1   As built drawings

Accepted: Date _____

6.2   As built dossier

Accepted   Date _____

6.3   Mechanical completion file

Accepted: Date _____

6.4   Spare parts lists

Received: Date _____

6.5   Start up, operating, and maintenance manuals

Accepted: Date _____

7.0   Cost and schedule comparison

7.1   Actual and planned progress comparison

A graphical presentation of the actual and planned progress curve together with a weekly tabulation of progress is inserted here.

7.2   Actual and planned man-hour comparison

Final manpower (man-hour) performance report is inserted here.

7.3   Actual and planned manpower comparison

A histogram showing actual and planned manpower is inserted here together with a weekly tabulation of manpower.

7.4   Contractor's cost management report

This report should include contractor's final cost in accordance with the contract terms if applicable.

8.0   Contractor's performance report

8.1   Management and adminstration

Engineering: _____

Project management: _____

Procurement and subcontracting: _____

Materials Management: _____

Construction management: _____

Project adminstration (including contract administration, cost control, scheduling, and accounts): _____

_____

Adequacy of facilities _____

Overall: _____

8.2   Construction labor and foremen

Attitude: _____

Workmanship: _____

Productivity: _____

Safety consciousness: _____

8.3   Cooperation with company: _____

8.4   Summary of strengths: _____

8.5   Summary of weaknesses: _____

_____

8.6   Problems arising through contractual terms, conditions, inadequate contract language, etc.

_____

_____

_____

9.0   Photographs: - attach any photographs made throughout the course of the project which may be useful as a record of progress or outstanding activities such as heavylifts, etc.

The contract closeout report may take some time for the contract engineer to complete, mainly due to the difficulty of obtaining input from the various site team disciplines involved. It is not uncommon for the contract engineer to fill in the gaps himself when he finds that a site superintendent who promised the information has disappeared on another project. Eventually, the work is done and all that remains is to distribute an interoffice memo for signature by the company representative to the effect that work has been completed under Contract No. _____ ; all monies owed and due to the contractor have been paid in full; and the contractor has furnished the company with a properly executed Certification and Release document to notify all interested parties that the contract is formally closed.

In some organizations, the contract engineer will also prepare a letter to the contractor reminding him that the work has been performed to the full written acceptance of the company and all outstanding financial obligations against the work have been paid in full. The contractor is notified that the contract is closed and no further work will be performed and no further payment will be made. The letter may express the company's appreciation for the manner in which the work in the contract was carried out and may assure the contractor that on future invitations to bid for work for which the contractor qualifies, every consideration will be given.

Model Contract Document

Standard Contract Forms

The contract document will consist of eight parts as follows:
Part 1. Articles of Agreement
Part 2. Scope of Work
Part 3. Master Schedule
Part 4. Payment details
Part 5. Material and equipment to be furnished by the client
Part 6. List of specifications
Part 7. List of drawing numbers
Part 8. Adjustments to Articles of Agreement (if required)

PART ONE

Articles of Agreement
1.   *Definitions*
       For the purpose of this Contract, the following titles shall be defined
as follows:
"Company" shall mean The Crude Oil Company Incorporate.
"Company Representative" shall mean the representative appointed from time to
time by Company in writing to act in such capacity for the purpose of this Contract.
"Company Furnished Items" shall mean the materials and equipment listed in Part
5. hereto.
"Contractor" shall mean Acclaim Contracting Incorporated.
"Contract Price" shall mean the agreed compensation specified in Part 4. of this
Contract, as such sum may be increased or decreased in accordance with this
Contract.
"Contract Schedule" shall mean the schedule for the Work as described in
Part 3. of this contract or as it may from time to time be amended in accordance
with this Contract.
"Contractor's Representative" shall mean the representative appointed from time
to time by Contractor in writing and approved in writing by Company's Represen-
tative to act in such capacity for the purposes of this Contract.

"Completion Certificate" shall mean the certificate issued by Company's Representative denoting completion of the Work as provided in these Articles.

"Drawings" shall mean the drawings referred to in Part 7. of this Contract. Such drawings may from time to time be supplemented or amended in accordance with the terms of this Contract.

"Facilities" shall mean all such property, services and equipment as may be needed by Contractor in order to perform the Work including but not limited to, the site, workshops, equipment, stores, offices, accommodations, and temporary works.

"Guarantee Period" shall mean the period described in these Articles.

"Quality Assurance Program" shall mean Contractor's written description of systems, plans, organization, job instructions, and control measures which will be used to ensure that the required quality will be achieved in performing the Work. The description shall relate to all aspects of Work execution including but not limited to, Work planning, Work management, Work instruction, quality control, inspection, testing, safety, and documentation.

"Site" shall mean the place or places where the Work or any part thereof as described in Part 2. of this Contract shall be performed.

"Specifications" shall mean the specifications referred to in Part 6. of this Contract. Such specifications may from time to time be supplemented or amended in accordance with the terms of this Contract.

"Subcontract" shall mean a contract entered into between Contractor and any person or company in the manner and to the extent permitted under the terms of this Contract.

"Work" shall mean all work to be performed in accordance with this Contract.

2.   *Company Obligations to Contractor*

        Company shall, in accordance with the terms and conditions of this Contract:

2.1   Make available to Contractor information necessary to the satisfactory performance of its work.

2.2   Provide permanent materials and equipment for the Work that are to be provided by Company in accordance with Part 5. of this Contract.

2.3   Obtain all permits, licenses, and other authorizations which must be obtained in Company's name and which are necessary for the performance of the Work.

2.4   Allow Contractor access, subject to normal, security and safety regulations, to the Work site as required for the performance of the Work.

2.5   Provide such security at the Work site as is to be provided by Company.

2.6   Appoint a Company Representative.

2.7   Perform all other obligations required of it by the terms of this Contract.

3.   *Contractor Obligations to Company*

        Contractor shall, in accordance with the terms and conditions of this Contract:

3.1   Carry out the Work in a professional and diligent manner to achieve the time frame and milestones specified in the Contract Schedule.

3.2   Provide permanent materials and equipment for the Work which are to be provided by Contractor in accordance with the Contract.

3.3   Obtain prior to the commencement of the Work and at Contractor's cost, all necessary information regarding taxes, laws and regulations and other conditions for carrying out the work.

3.4   Obtain all permits, licenses, and other authorizations which are necessary for the performance of the Work except those which must be obtained in Company's name.

3.5   Provide support facilities, construction plant, and each and every item of tools and equipment as required by this Contract.

3.6   Provide all of the labor and supervision required to perform the Work.

3.7   Provide such security at the Work site as is to be provided by Contractor.

3.8   Provide "as built" drawings to facilitate operation and maintenance of the completed Work.

3.9   Appoint a Contractor's Representative for the duration of the Work.

3.10   Perform all other obligations, work and services as required by the terms of this Contract.

4.   *Contractor Personnel*

4.1   Contractor shall provide sufficient competent and fully qualified supervisory personnel to execute the Work in the manner and within the time required by this Contract.

4.2   Upon Company's written request, Contractor shall, at its own cost and expense, remove from employment in the Work any Contractor personnel determined unsuitable by Company.

4.3   Contractor personnel assigned for the performance of the Work and approved by Company shall not be removed or transferred by Contractor without the prior approval of the Company Representative.

5.   *Contractor Site Investigation*

5.1   Contractor agrees that it has thoroughly investigated and satisfied itself as to all the general and local conditions affecting the Work or has had the opportunity to do so, including but not limited to, transportation and access to the Work site including the availability and conditions of roads, disposal, handling and storage of materials, availability and quality of labor, water, and electricity; climatic conditions, tides, and ground water; and equipment, machinery, and materials required by Contractor prior to and during performance of the Work.

5.2   Contractor agrees that it has thoroughly investigated and satisfied itself as to the condition of the Work site and approaches to the site including but not limited to, topography and ground surface conditions, subsurface geology and the nature of surface and subsurface materials and obstacles to be encountered such as buried pipelines, in service or abandoned, drainage systems, and the like.

6.   *Duration of Contract and Work Schedule*

6.1   Contractor shall commence performance of the Work on the appropriate date set forth in the Contract Schedule and shall continue the same in an expeditious manner and in accordance with the completion date and the milestones set forth in the Contract Schedule. Should the Work extend beyond the scheduled completion date for whatever reason, the Contract will remain in full force and effect until termination under the terms of the Contract.

6.2   Contractor acknowledges and agrees that actual delays in activities which, according to the base line schedule, do not effect any milestone or completion dates shown by the critical path in the schedule, do not have any effect on the contract completion date or dates and therefore will not be a basis for a change therein.

6.3   The Work shall be considered to have reached "mechanical completion" when the permanent works described in the Contract Scope of Work, or a portion or portions thereof, have been mechanically and structurally put in a tight and clean condition and otherwise constructed as provided in the Contract and all deficiencies, including those which could prevent or delay safe and orderly prestart-up or start-up procedures by Company or timely achievement of permanent works operation at the conditions specified in the Contract, to the extent that such deficiencies may be determined, have been corrected.

6.4   Contractor shall notify Company Representative of possible delays to the schedule or achievement of milestone date or dates for whatever reason. Notification must be in writing and made within 24 hours of the date Contractor first had cause to believe that the performance of the Work may be delayed. Such notification will include:

(a)   the amount of delay Contractor considers will or could be incurred and

(b)   any remedial action Contractor proposes to adopt to avoid delay.

6.5   If at any time during the performance of the Work, Company considers that Contractor's progress is insufficient to meet or keep pace with the schedule, Company may instruct Contractor to take the necessary action to improve its progress. Should Contractor's performance not improve within a reasonable period following this instruction, Company may instruct Contractor to increase its labor force, number of shifts, introduce overtime work, and/or additional days of work per week at Contractor's expense.

7.   *Variations to the Work*

7.1   Company has the option at any time to make changes within the general scope of this Contract including but not limited to, changes in the sequence of performance of the Work, alterations to or in the design of the permanent Work or increase or decrease the quantity of the Work and Contractor shall perform the Work as changed.

7.2   Changes to the Scope of Work shall be made in writing by means of a Work Order where there is a fixed price or agreed unit price in Part 4. of this Contract and in accordance with the conditions governing the issue of a Work Order as described in Procedure Number *** which, by this reference is made part of this Contract.

7.3   Changes to the Scope of Work shall be made in writing by means of a Change Order where there is no fixed price or agreed unit price in Part 4. of this Contract and in accordance with the conditions governing the issue of a Change Order as described in Procedure Number *** which, by this reference, is made part of this Contract.

7.4   Where a change does not affect the Scope of Work or affects the Scope of Work but of sufficient substance to warrant a fully executed codicil to the Contract or changes the Articles of Agreement after the initial execution of the Contract, the change or addition shall be made in writing by Contract Amendment in accordance with the conditions governing the issue of a Contract Amendment as described in Procedure Number *** which, by this reference, is made part of this Contract.

7.5   All Contractor's obligations including guarantees contained in this Contract shall become applicable to each and every variation.

7.6   Any variation made necessary directly or indirectly by any error or omission of Contractor shall not constitute a change in the Work and shall not affect the Contract Price or Contract Schedule.

8.   *Guarantees and Remedy of Defects*

8.1   Contractor warrants that it shall perform the Work in a professional and workmanlike manner and that the permanent works as constructed shall conform to the final Drawings and Specifications and the permanent works shall be free of defects and suitable for the purposes intended.

8.2   If at any time prior to the Final Acceptance of the permanent works by Company or within the period specified in Part 2. of this Contract thereafter Company discovers that the permanent works as constructed by Contractor do not conform to this warranty, Contractor shall, after receipt of notice in writing from Company, promptly perform or arrange for the performance of any remedial work required to make the permanent work conform to this warranty. Such remedial work shall be performed to Company's satisfaction and at Contractor's expense.

9.   *Drawings and Specifications*

9.1   Contractor shall keep at the Work site a copy of all drawings and specifications as amended and revised from time to time. Anything mentioned in specifications and not shown on drawings or shown on drawings and not mentioned in specifications, shall be of like effect as if shown or mentioned in both.

9.2   During the performance of the Work, Contractor shall maintain a set of construction drawings to reflect the current "as built" status of the Work. Contractor shall submit to Company one set of final transparencies redrafted to reflect all changes to the permanent work prior to the submission of Contractor's final invoice to Company.

10.   *Inspections and Tests*

10.1   Contractor shall be responsible for the inspection and testing of all materials and workmanship to be provided by Contractor or its Subcontractors under the terms of this Contract.

10.2   Inspection or omitting to inspect by Company shall not be considered by Contractor as acceptance of any part of the Work.

10.3   Contractor shall provide, as required, all facilities, labor, and materials necessary for inspections or tests by Company at Contractor's expense.

10.4   Contractor shall, at its expense and when requested by Company's Representative, provide testing samples to be used in the Work.

10.5   No part of the Work shall be covered up or put out of view without the approval of Company's Representative. Contractor shall give sufficient advance notice and full opportunity for Company's Representative to inspect and test any part of the Work that is about to be covered up or put out of view.

11.   *Liability and Insurance*

11.1   Subject to the terms and conditions of this Contract, Contractor shall defend, indemnify, and hold Company harmless from any and all claims, losses, expenses, costs, or damages arising from or related to the injury to or death of any person and the damage to or loss of any property resulting from any and all acts or omissions of Contractor, its Subcontractors, or the personnel or agents of any of them.

11.2   Company shall release, indemnify, and hold Contractor harmless from all claims by Company for injuries or death of employees of Company or loss or damage to Company property or the permanent works to the extent that such claims, losses, expenses, costs, or damage are not recoverable under insurance policies purchased by Contractor in accordance with the requirements of this Contract.

11.3   Contractor shall procure and maintain at all times during the term of this Contract the insurance policies described below:

11.3.1 Such Workmens' Compensation and Employer's Liability Insurance as shall be necessary and adequate to cover all Contractor's personnel while engaged in the Work.

11.3.2 Comprehensive Bodily Injury and Property Damage Liability Insurance covering all Contractor's work under this Contract.

11.3.3 Public Liability and Property Damage Insurance covering Contractor's owned or hired vehicles and equipment.

12.   *Suspension of Work*

Company may at any time, with or without cause, suspend performance of all or any part of the Work by giving Contractor prior notice in writing specifying the Work to be suspended and the effective date of such suspension. Such suspension may continue for a period of up to *** calendar days after the effective date of suspension during which period Company may direct, in writing, Contractor to resume performance of the Work. Contractor shall take all actions necessary to maintain and safeguard the suspended Work. Company shall not be liable for loss of anticipated profits or for any damages or any other costs incurred with respect to suspended Work during the period of suspension, except for reasonable and auditable costs which:

(a)   are incurred for the purpose of safeguarding the Work and materials and materiel in transit to or at the Work site;

(b)   are incurred for such Contractor or Subcontractor personnel or equipment which Contractor continues to maintain at Company's request, at the Work site; or

(c)   are otherwise reasonable and unavoidable costs of suspending the Work and of reassembling personnel and equipment.

If at the end of the ** calendar days period Company has not required a resumption of the Work, that portion of the Work that has been suspended shall be deemed terminated as of the effective date of suspension unless Company and Contractor shall agree to a further extension of the suspension period.

13.   *Termination at Company Option*

13.1   Company may terminate this Contract or any part of the Work by giving notice in writing to Contractor specifying the Work to be terminated and the effective date of termination.

13.2   Should Company give notice to terminate this Contract or any portion of the Work, Contractor shall stop performance of the Work involved on the effective date of termination. Company shall pay Contractor all amounts properly due up to that date. Additionally, Company shall pay Contractor auditable costs incurred as a direct result of such termination including but not limited to, reasonable cancellation charges paid by Contractor to its Subcontractor or vendors, reasonable demobilization charges, and reasonable costs incurred in preserving or protecting materials and/or materiel or Work in progress at the time of termination, plus an amount equal to **% of the foregoing termination costs.

14.   *Termination for Default*

14.1   Should Contractor commit a substantial breach of this Contract, Company may demand, in writing, that Contractor comply with the terms of this Contract. If within ** days after receipt of such a demand, Contractor has not complied or has failed to take satisfactory action to comply or within ** days Contractor has not remedied the breach, Company may terminate this Contract by giving Contractor written notice to that effect.

14.2   Should Contractor commit an act of bankruptcy or seek legal or equitable relief for reasons of insolvency or become unable to meet its financial obligations, Company may terminate this Contract by giving Contractor written notice to that effect.

14.3   On the day on which termination becomes effective, Contractor shall stop performance of the Work. Company shall be entitled to perform or cause to be performed by third parties, the balance of the Work remaining. All costs thereof shall be borne by Contractor and shall be recoverable from Contractor by any means available to Company under the terms of this Contract or at law, including without limitation, the deduction by Company of any such costs from any monies due or that may become due to Contractor.

15.   *Force Majeure*

15.1   Neither Company nor Contractor shall be considered in default in the performance of their obligations to the extent that the performance of such obligations has been delayed, hindered, or prevented by force majeure. The term "force majeure" as used in this contract shall mean, cover, and include the following: Acts of God, acts or restraints of governmental authorities, fire, explosions, storms, wars, hostilities, blockades, public disorders, quarantine restrictions, embargoes, strikes or other disturbances, loss or shortage of transportation facilities, breakdown of machinery and equipment not caused by the negligence of the party rendered unable to perform its obligations, or any other act, event, cause, or occurrence rendering a party unable to perform its obligations which is not in the reasonable control of such party, whether or not similar to any of the foregoing.

15.2   If an event of force majeure should occur, the party affected shall notify the other party of such occurrence within ** days thereof. Should such event continue without interruption for a period of *** days or more, either party shall have the right to terminate this Contract by giving at least ** days notice in writing to the other party specifying the date upon which such termination shall occur and provided that the state of force majeure does not come to an end before the date specified for termination in said notice, this Contract shall terminate on the date so specified.

16.   *Contract Price and Payment*

16.1   As full and complete compensation for Contractor's performance of the Work and all of Contractor's obligations hereunder in accordance with the terms and conditions of this Contract, Company shall pay Contractor the price set forth in Part 4. of this Contract.

16.2   The Contract Price shall be increased or decreased only by Amendment. When the total value of current Work Orders and/or Change Orders reaches **%

of the value of the Contract as amended, a Contract Amendment shall be raised to adjust the Contract Price.

16.3   Payment of the Contract Price to Contractor shall be made as provided in Part 4. of this Contract.

Note 1.

This model Articles of Agreement is offered as a foundation on which the panel of experts may build a standard document. It is obviously not complete since many of the missing clauses will relate to different countries, laws, and customs. It is also intended for use with Lump Sum contracts or without unit rates for extra work. Further clauses would need to be added to accommodate reimbursable cost contracts.

Note 2.

Space for the addresses and signatures of client and contractor is usually reserved at the last page of the Articles of Agreement.

PART TWO – THE SCOPE OF WORK

1.   *Scope of Work – Summary*

Except as otherwise expressly mentioned in this Contract, Contractor shall provide all labor, supervision, engineering and services, quality control, inspection and testing, transportation, installed and consumable materials, equipment, tools, storage, construction camp accommodation, and each and every item of expense necessary for the supply, fabrication, erection, installation, application, handling, hauling, unloading and receiving, construction, evaluation, design engineering, testing, assembly, and production of _____ hereinafter called the Work.

2.   *General Description of Permanent Works*

(Here will follow a brief description of the facilities the client proposes to build in the overall Project. This section will describe all of the new structure even though the contractor to whom this contract refers is only concerned with a portion of the Work.)

3.   *Scope of Work – Detailed*

(The detailed Scope of Work must be tailored to each specific project but must be sufficiently comprehensive to cover every aspect of the proposed work.)

PART THREE – THE CONTRACT SCHEDULE

1.   The Permanent Works shall be Mechanically Complete by the Scheduled Completion date of _____ 199 ___

Critical Milestone dates shall be completed as specified below:

Plant A _____ 1 January   199 ___
Plant B _____ 1 March   199 ___
Plant C _____ 1 August   199 ___
Plant D _____ 1 December 199 ___

PART FOUR – THE CONTRACT PRICE AND PAYMENT

1.   *Contract Price*
1.1   As full and complete compensation for Contractor's performance of the Work and of all of Contractor's obligations in accordance with the terms and conditions of this Contract, Company shall pay Contractor a Lump Sum Contract Price of _____ .

1.2   Except as provided hereunder, the Contract Price constitutes the entire compensation due Contractor for the Work and all Contractor's obligations in accordance with the terms and conditions of this Contract. The Contract Price includes but is not limited to, compensation for all applicable taxes, fees, overheads, profit, mobilization and demobilization, and all other direct and indirect costs and expenses incurred or to be incurred by Contractor. The Contract Price includes for all overtime premiums and payments when such overtime is necessary to meet the schedules completion date and critical milestone dates.

2.   *Variations*
Compensation for authorized changes to the Work shall be made as follows:
2.1   Lump Sum basis
2.2   Unit Rate basis as set forth in Attachment 1. to this Part Four
2.3   Labor and equipment rate basis as set forth in Attachment 2. to this Part Four.

Note 1. Equipment rates will constitute all inclusive payment to Contractor per hour contractor's equipment is operable and available. These rates shall apply to actual hours that equipment is operating or available at the work site for use at any time of the day or night or any day of the week. They are inclusive of mobilization and demobilization, labor, and costs for all maintenance, repairs, refueling and lubrication, spare parts, tools, consumables, depreciation, insurance and taxes, and all overheads and profit. They are inclusive of the cost of the operators.

Note 2. Labor rates will constitute all inclusive payment to Contractor and will include for the following:
Wages, piecework, other allowances (traveling, and the like), all statutory and other related charges such as National Insurance, Graduated Pensions, Insurances, Company Pensions, Holiday Pay, Sick Pay, Workmens' Compensation and Employer's Liability Insurances, Industrial Training Levy, Redundancy Payments Contributions and any other similar charges, small tools such as ladders, chisels, handsaws, buckets, hammers, hard hats, protective clothing, consumables, gases, welding rods, oils, lubricants, electricity, all maintenance charges on facilities, all supervision and management required, all administration and establishment charges, overheads and profits, all necessary engineering services, all temporary facilities, personnel transport, and all delay and disruption caused as a result of performing work on an hourly rates basis.

3.   *Payment*
3.1   Company shall make payment against Contractor's invoices for actual progress of the Work performed in accordance with the provisions of Article * * of this Contract.

Note: The above mentioned clause is not included in the model Articles of Agreement in Part 1. as the terms will have to be fashioned to suit the client requirements in each separate project. For details of various payment arrangements, see Chapter Eight.

4.1   Retention
     The retention amount which will be withheld by Company from payments due to Contractor in accordance with Article ** of this Contract, will be ** percent. Alternatively, Company may agree to an arrangement whereby Contractor may provide an irrevocable bank guarantee in a form acceptable to Company, in lieu of retention, for ** percent of the cumulative amount to be paid by Company against Contractor's invoices over the duration of the Contract.
5.   *Pro Forma Invoice*
     Contractor shall submit a Pro Forma invoice to Company Representative ** working days prior to Contractor's submittal of any progress invoice under this Contract. The "Pro Forms" invoice shall be clearly marked "not an application for payment" and will be complete with all relevant supporting documentation in order that Company and Contractor shall have a period of review of Contractor's claim for payment and agree or adjust any contentious items prior to formal invoice submissions.

PART FIVE – COMPANY SUPPLIED MATERIALS AND EQUIPMENT

Note: During the planning of a very large project, most clients will elect to purchase and supply to the contractors material and equipment to be used in the permanent works or will have the managing contractor do so. The client's organization is usually better placed to procure, inspect, and expedite large amounts of material and materiel and to arrange shipping and customs clearance to site. A major oil company would probably have an economical advantage over a single contractor although it is not unusual for the client to arrange bulk procurement of all permanent materials and negotiate the financial terms leaving the contractor to find the actual funds to pay for the goods and bring them to the site as part of the contract deal.
Where the client is to furnish the materials and equipment for the permanent works and deliver to site, this Part 5. would list the items to be supplied and may also record the expected delivery dates.

PART SIX – LIST OF SPECIFICATIONS

Note: This Part 6. will contain only the reference numbers and titles of the specifications to be used on the project. The actual specification documents will be delivered to the contractor before commencement of the work.

PART SEVEN – LIST OF DRAWING NUMBERS

Note: This Part 7. will record the drawing numbers and titles of all drawings to be used on the project. A large project may issue over 100,000 drawings including vendor drawings and the total will almost certainly not be delivered to the contractor before start-up.

PART EIGHT – SPECIAL TERMS AND CONDITIONS

Note: This Part 8. is only used where it is required to deviate from the preprinted Articles of Agreement or to introduce special conditions not provided for in the other parts of the contract.
Not all construction contracts require a Part 8. particularly where adjustments, even to the Articles, are made before execution. However, once a suitable model is made for the purpose of standardization and the Articles are separately printed, it would seem advisable to make adjustments through an insertion into Part 8. on the rare occasions when this becomes necessary. This is particularly appropriate where the aim is to produce a model Articles of Agreement for international acceptance and use.

Amendments

Amendment to Contract

Contract No. 123456   Amendment No. 002

THIS AMENDMENT is entered into, effective as of the day of _____ 19___ by and between

>THE CRUDE OIL COMPANY
>having its registered office at
>Shale Beach, San Francisco,
>California, U.S.A.
>(hereinafter referred to as Company)

and

>ACCLAIM CONTRACTING INCORPORATED
>whose address is
>1 Sandy Lane, Los Angeles,
>California, U.S.A.
>(hereinafter referred to as Contractor)

In consideration of the agreements herein contained, the parties hereto agree as follows:

1.0   *Amendments*

The Contract heretofore entered into by the parties dated effective as of the day of _____ 19___ and identified by the Contract number set forth above, hereinafter referred to as Contract, is hereby further amended as follows:

1.1   *Part I – Articles of Agreement*

1.1.1 Article 16. *PATENT DESIGN AND COPYRIGHT INDEMNIFICATION* Subsection 16.1

Delete this Subsection in its entirety.

1.2   *Part II – Scope of Work*

1.2.1 Section 3.0 *Welding* Subsection 3.4

Change the first sentence to read:

"Contractor shall qualify the welding procedure to be used within thirty (30) days of the award of the Contract."

1.3   *Part III – Contract Schedule*

183

1.3.1 Section 2.0 *Contract Completion*

Change the Contract completion date from 1 March 1991 to 1 February 1991.

1.4   *Part IV – Compensation*

1.4.1 In consideration of the change described in 1.3.1 above, Contractor shall be paid an acceleration fee of Twenty thousand U.S. Dollars ($20,000.00) upon the achievement of mechanical completion as defined in the Contract by the amended date of 1 February 1991.

2.0   *Pricing Summary*

2.1   The Contract Price set forth in the Contract is hereby changed as follows as full compensation to Contractor for full and complete performance by Contractor of this Amendment in full compliance with all terms and conditions of the Contract.

Total net addition to Contract Price:
$20,000.00

2.2   Original Contract Price: $1,000,000.00
       Net adjustment of Contract Price pursuant
       to previous Amendments: $ 500,000.00

2.3   Net adjustment of Contract Price pursuant
       to this Amendment: $ 20,000.00

2.4   Total Contract Price: $1,520,000.00

3.0   *Status of Contract*

As Amended herein, the Contract shall continue in full force and effect,

IN WITNESS WHEREOF, the parties hereto have executed this Amendment on the day and year below written but effective as of the day and year first set forth above.

ACCLAIM CONTRACTING INCORPORATED

By _____

Title _____

Date _____

THE CRUDE OIL COMPANY

By _____

Title _____

Date _____

Note that the Amendment, similar to most contracts, is not necessarily made effective on the same day that the document is signed. Unlike the change order, an amendment does not have to be fully executed before work relating to the amendment may commence, although the period between the effective date and the signature dates should not be too protracted.

It will be seen that our sample amendment is the second one issued on our fictitious contract and some clients, under Section 1.0, would draw attention to this by writing,

"The Contract heretofore entered into by the parties dated effective as of the _____ day of _____ 19__ and identified by the Contract number set forth above, *as previously amended by Amendment number(s)* _____ *thereto,* hereinafter referred to as "Contract" is further amended as follows:"

It is usual practice to have the contractor sign first for both amendments and change orders although the practicality of the latter method is debatable since the change order is, in effect, an instruction from the client to carry out extra work and one would imagine that the client should sign first, but no one seems to raise much objection.

Model Change Order

Contract Change Order

Date: _____   Contract No.: _____

Change Order No: _____

The following changes shall be performed by Contractor before _____

19__

Payment for all work described in this Change Order shall be

(a) $ _____ Lump Sum (b) Estimated at $ _____

(c) Shall not exceed $ _____

The payment described above shall include Contractor's profit and shall be in full and final settlement of all costs, expenses, and overheads, whether direct or indirect, incurred by Contractor and arising from or related to but not limited to: (1) the extra work described in this Change Order; (2) any delay caused to the schedule or any change in the completion date of the Project resulting from said extra work; (3) the cumulative effect, if any, of said extra work in conjunction with extra work described in any other Change Orders.

Description of Work _____

_____

_____

_____

Backup information attached _____

_____

Change proposal initiated by: _____

Justification for the Work _____

Anticipated completion date _____

Effect on baseline schedule _____

Cost Code(s) _____

Increase in Contract Price or Decrease (Negative Change Order) _____

_____

Company   Date: Contractor   Date:

Model Work Order

Contract Work Order

Date: _____   Contract No.: _____

                                    Work Order No: _____

The following changes shall be performed by Contractor before _____
19__
Payment for all work described in this Work Order shall be $ _____
This sum includes Contractor's profit and shall be in full and final settlement of all costs, expenses, and overheads, whether direct or indirect, incurred by Contractor and arising from or related to but not limited to: (1) the extra work described in this Work Order; (2) any delay caused to the schedule or any change in the completion date of the Project resulting from the said extra work; (3) the cumulative effect, if any, of said extra work in conjunction with extra work described in any other Work Orders.

Description of Work _____
_____
_____
_____
_____

Backup information attached _____
_____

Work proposal initiated by: _____

Justification for the Work _____
_____

Anticipated completion date _____

Effect on baseline schedule _____

Cost code(s) _____

Company   Date: Contractor   Date:

Model Short Form Contract

(For lump sum or unit rate work valued at less than $50,000.)

THIS CONTRACT IS entered into, effective as of _____ by and between _____ (hereinafter referred to as "Company"), whose address is _____ and (hereinafter referred to as "Contractor"), whose address is _____

In consideration of the agreements herein contained, the parties hereto contract and agree as follows:

PART 1 - GENERAL TERMS

Article 1.0 Contract Documents
This Contract shall consist of the following Parts:
Part 1 - General terms of Agreement
Part 2 - Scope of Work
Part 3 - Schedule
Part 4 - Commercial terms and the exhibits, drawings, specifications and documents referred to therein, all of which, by this reference are made part of this Short Form Contract which sets forth the entire agreement between the parties and supersedes all previous communications, agreements, and commitments, whether written or oral.

Article 2.0 Variations
No variation orders or amendments shall be made to this Contract. Should the Scope of Work require changes, this Contract shall be canceled and a replacement Contract shall be executed incorporating the change or changes.

Article 3.0 Insurance
3.1 Contractor shall, at its own expense, maintain during the entire progress of the Work, insurances of the following descriptions and limits:
3.2 Workers' Compensation in accordance with the provisions of the applicable Workers' Compensation law or similar laws of the state, territory, province, or division having jurisdiction over the employee and Employer's Liability with a limit of liability of $XXX,XXX for each occurrence.

3.3 Automotive Liability covering use of all owned and hired vehicles with a combined single limit of liability of $XXX,XXX per occurrence for bodily injury and property damage.

3.4 Certificates of insurance satisfactory to Company shall be supplied by Contractor confirming that the above insurances are in force.

Article 4.0 Work Rules, Laws, and Regulations

4.1 Contractor shall comply with Company's work rules and with all local, municipal, state, federal and governmental laws, orders, codes, and regulations applicable to Contractor's operations in the performance of the Work and shall obtain and comply with, at its own expense, all permits, certificates, and licenses required of it by governmental authority.

Article 5.0 Guarantees

Contractor guarantees Company that the Work shall strictly comply with the provisions of this Short Form Contract and all drawings and specifications relevant to this Contract and shall be free from defects in design, materials, construction, and workmanship.

Article 6.0 Cleanup

Contractor shall keep the work site and the vicinity thereof clean and free from any debris and rubbish caused by the Work and on completion of the Work shall leave such areas clean and ready for use.

PART 2 - SCOPE OF WORK

2.1 Except as otherwise expressly provided in this Short Form Contract, Contractor shall supply all labor, supervision, installed and consumable materials, equipment, tools, consultation, services, testing devices, storage, and each and every item of expense necessary for the supply, fabrication, erection, installation, application, handling, hauling, unloading and receiving, construction, evaluation, design engineering, testing, assembly, and production of _____ hereinafter called the Work.

$\left\{ \begin{array}{l} \textit{insert detailed scope of work including lists} \\ \textit{of drawings and specifications} \end{array} \right\}$

PART 3 - SCHEDULE

3.1 Work shall be completed by _____

3.2 Company shall schedule and coordinate Contractor's performance of the Work with the work of others and Contractor agrees to comply strictly with such scheduling and coordination. Specific scheduling requirements are _____

_____

_____

PART 4 - COMMERCIAL TERMS

4.1 Full compensation to Contractor for complete compliance with all provisions of this Short Form Contract shall be as stated below. This compensation shall be firm for the duration of the Work and shall include all Contractor's costs, taxes, duties, license fees, expenses, overhead, and profit.

No progress payments will be made and Contractor shall submit a single invoice for the Work to the total value of this Short Form Contract following completion and acceptance of the Work.

$ _____

IN WITNESS WHEREOF, the parties hereto have executed this Short Form Contract on the day and year below written but effective as of the day and year first set forth above.

COMPANY _____   CONTRACTOR _____

By _____   By _____

Title _____   Title _____

Date _____   Date _____

Glossary of Terms

This glossary is a collection of alphabetically listed terms and definitions commonly used in the oil, gas, and petrochemical construction business. These may include legal terms, terms used in standard contract documents, insurance terms, common shipping terms, or general terms. To obtain uniformity in the usage of these terms, the definitions found in this glossary shall be deemed to be the author's definitions for contracts management purposes.

ADDENDUM–Any written alteration or addition to an existing, published, and released Request for Quotation, Invitation to Bid, Bid Package, and the like.

ADVANCE MOBILIZATION PAYMENT–When considering a contract in which the contractor has massive front end expenses in the form of establishment of camps, purchase of equipment especially for the work, and the like, the client may agree to make an advance mobilization payment free of interest or with low interest payable through invoice deductions over the life of the contract. This arrangement may be announced in the bid package so that the bidders may be competitive in their bids, always assuming that the client is able to obtain money cheaper than the bidders. There would normally be a requirement by the client for a bank guarantee to cover this loan.

AMENDMENT–A fully executed, written alteration to an existing contract. The Contract Amendment will be used rather than any other form of variation order when the alteration does not fall within the general scope of the contracted work. The Amendment is signed by both parties as a codicil to the contract, preferably by the original signatories of the contract. The Contract Articles of Agreement, for example, should only be changed by Amendment. A further example is in the case of the contractor changing its company title during

the course of the work. More seriously, a very substantial alteration to the compensation would merit an Amendment. A more
detailed treatment of this subject is found in Chapter 6 of this book.

ARBITRATION–In the comparatively rare situation where the
contractor rejects the client's final determination on a claim, the
contract will make provision for the dispute to be settled in court
or by reference to arbitration. If the latter course is available in
the contract (not all contracts carry an arbitration clause), both
parties will each choose a person to act a arbitrator and usually,
a third arbitrator is appointed by the two chosen arbitrators. The
arbitrators so appointed will hear the dispute and will base their
decision on the rules of the law and the terms of the contractual
arrangement between the parties. For their part, the parties will agree
that the award of the arbitrators shall be final and binding upon
the parties.

ARTICLES OF AGREEMENT–The "preprinted" part of the contract, collated by the client's legal department and appearing,
normally unaltered, in every one of that client's contracts. The
Articles cover definitions of terms used in the contract, variations,
guarantees, audit rights, subcontracts, indemnity and liabilities,
insurance requirements, default, force majeure, suspension, and
termination. There is usually a space at the end of the Articles for
the official signatures and seals of the contracting parties. In most
client organizations, the Articles may only be altered by Amendment and then only with prior agreement from the legal department.

"AS BUILT" DOCUMENTATION–A complete record of the final
situation of the work on completion; handed to the client by the
contractor in the form of marked up drawings, alignment sheets,
test reports, certifications, and other document necessary for the
operation of the work (see Chapter 3).

AUTHORIZED REPRESENTATIVE–A client representative or a
contractor representative appointed in writing by the respective
parties to direct the contracted work.

BACKCHARGE–A charge against contractor or manufacturer for costs
incurred by the client as a result of unilateral corrective action due
to unacceptable performance by the contractor, or unacceptable standards of material by the manufacturer; or predetermined charges for
equipment, services, or materials provided by the client to the contractor during the performance of its contract which are incurred by
the client on behalf of the contractor or the contractor's subcontractor.

BANK GUARANTEE–A client may accept an irrevocable bank guarantee in lieu of retention, thus saving the contractor's funds which would otherwise be tied up but at the same time, placing money within reach of the client in the event of default.

BENEFICIAL OCCUPANCY–The advantageous possession and utilization of the work by the client following completion or at a stage (substantial completion) when the building or facility may be used for its intended purpose.

BIDDER–A contractor submitting a proposal in response to a Request for Quotation.

BID EVALUATION–An assessment of completeness and overall acceptability of the bids received for a given contract including an evaluation and comparison between the contractors for cost, schedule, reliability, and ability to perform in accordance with the Request for Quotation.

BID SUMMARY–A record of all bids received and evaluations made on such bids.

BILL OF QUANTITIES–Usually concerned with the civil work in a contract, the Bill of Quantities is a list of items in the contractor's work with columns for the estimated quantity per unity of measure i.e., linear feet, cubic yards, and so on), the contractor's quoted unit price and the sum.

BONDS

BID BOND–A type of surety bond that is submitted with a contractor's bid for work. It demonstrates good faith and protects the client against a loss incurred when the selected bidder fails to accept the contract award. The bid bond may stipulate that a performance bond and labor and material bond will be provided upon award of the contract.

LABOR AND MATERIAL BOND (Payment Bond)–A type of surety bond that protects the client against the contractor's default in the supply of material, equipment, or labor.

PERFORMANCE BOND–A type of surety bond designed to protect the client against loss caused by acts or failures of a contractor in the performance of its contract.

SURETY BOND–An agreement by one called a surety (guarantor), to assume the debt, default, or miscarriage of another called the principal. The primary types of surety bonds are the bid (proposal) bond, performance bond, and the labor and material (payment) bond.

BREACH OF CONTRACT–Violation of agreement between contractor and client.

CASH FLOW PROFILE–A chart showing expenditure on a project and income from the project over a period of time.

CERTIFICATE OF INSURANCE–The document used as evidence of insurance in lieu of a policy. It includes a statement of coverage in general terms and the limits of liability for each category of coverage shown. The Contract Articles of Agreement oblige the contractor to furnish to the client a certificate of insurance listing all policies carried by itself and its subcontractors. This proof of cover must be produced prior to work affecting such policies.

CERTIFICATION AND RELEASE–The document that some clients use to effect a lien release. It is most often used when closing out a contract. It should be notarized and transmitted with the contractor's request for final payment.

CHANGE ORDER–A formal document, issued by the client, to authorize a change in the contract scope of work.

CHANGE ORDER CLOSEOUT FORM–A simple form signed by the client and the contractor, closing out each completed change order with the final amount to be paid.

CHANGED CONDITIONS–Unscheduled and /or unexpected deviation in the job site or contract conditions.

CHANGED WORK–That work which results from nonquantitative additions and deletions and/or modifications to the work contemplated by the contract which may result from a design change or some field requirement. It can be but is not necessarily a scope change. It may cost money or result in price reduction.

CLAIM–In this context, a claim is a unilateral action by a contractor wherein he claims extra compensation or time based upon his own evaluation of a situation in relation to his own interpretation of the contract. When a contractor considers that he is in a claim situation, he will advise the client accordingly, with a view to receiving compensation often in the form of a change order. If the client does not reject all of the contractor's claim, the indisputed portion may be settled in this manner. The balance of the claim may be debated or negotiated or, in extreme cases, referred to arbitration or the law courts.

CLARIFICATION MEETING–A meeting held with the bidders during the bid evaluation period the purpose of which is to exchange information, clarify the bidders' intentions, and explain the modus operandi regarding bidders' offers.

CLIENT–Also OWNER, COMPANY, or the abbreviated title such as Mobil, Exxon, Shell, and the like. The party with whom the contractor makes an agreement or contract for the performance of work. In some superprojects, the client may appoint a managing contractor to control the work and issue contracts in the client's name or in the managing contractor's name on behalf of the client.

CODE OF ETHICS–An accepted system of principles and professional standards of conduct in the performance of business activities.

COMPANY ESTIMATE–Client appraisal of what it would cost the average, efficient contractor to perform the work, plus a reasonable profit.

COMPANY REPRESENTATIVE–Duly authorized representative of the client as defined in the contract.

CONFLICT OF INTEREST–When involving the contracts engineer, this amounts to a betrayal of trust and the abandonment of responsibilities in return for the promotion of private gain.

CONSTRUCTION CONTROL PLAN–A manual developed in house by the client's construction department for the guidance of its supervisory staff in the field.

CONTRACT–An agreement between the client and another party (the contractor) for the performance of work through the supply of labor and materials.

CONTRACT DOCUMENTS–As a minimum, these will consist of the contract as mentioned above, the specifications and the drawing, plus any other documents or references that are specifically described in the contract.

CONTRACTOR–An individual, firm, corporation, or other legally recognized entity performing work under the terms and conditions set forth in a contract. In most contracts and other legal documents, the contractor is referred to "it" and never as "he" or "him", e.g., "Contractor shall at all times keep its work area in a neat, clean, and safe condition." However, for the purpose of this work, the contractor will be referred to in the third person.

CONTROL INSTITUTION–An organization or company of high repute used by the client to perform specially defined approvals and verification work including, in some cases, laboratory testing and reporting. Welding procedures, for example, may be witnessed by such a body during the contractor's initial production. American Bureau of Shipping, Bureau Veritas, and Lloyds are companies specializing in this work.

COST CONTROL–As practiced by client's cost engineers, provides the project management with ongoing analyses of costs for individual contracts and for the project as a whole. The principles of cost control are:

Estimating. In the very early pre-engineering stages of a project, preliminary control estimates of probable cost are made from conceptual data. These lead to the master control estimates at the end of the pre-engineering phase, which include base estimates from more detailed engineering information plus escalations and contingency appraisals. In the construction stage, current control estimates are used to provide information from ongoing bid evaluations, contract awards, and forecasts.

Monitoring means checking closely on contracted work in progress and watching for deviations from the original plan.

Analyzing. The cost engineer, with construction staff and the contracts engineer discusses this information and uses it to make forecasts.

Forecasting. The probable end result is predicted, enabling management to take the appropriate corrective action.

Reporting. Regular and timely reporting makes client management aware of events as they arise.

COST AND FREIGHT (C & F)–The type of shipping contract in which the seller provides the product and the vessel and delivers the product to the nominated discharge port.

COST, INSURANCE, AND FREIGHT (CIF)–The type of shipping contract in which the seller provides the product and the vessel, procures the insurance, and delivers the product to the nominated discharge port.

COST PLUS–See REIMBURSABLE CONTRACT.

DAY RATES–These are related to the reimbursable cost contract. Day rates are utilized for equipment rental where the work is nor clearly defined or carries a degree of risk. They are often used when contractors provide specified equipment or personnel services such as offshore pile driving barges or diving services.

DELAY AND DISRUPTION–Interruptions to a contractor's performance caused by indirect impacts arising from direct impacts associated with unscheduled events such as scope of work changes that are not part of the contractor's planned management and work processes (see Chapter 7).

DOCUMENT AUDIT–In project management terms, this is a documented activity performed in accordance with written procedures or checklists to verify, by examination and evaluation of objective evidence that applicable elements of the Quality Program have been developed, documented, and implemented in accordance with specified requirements. A documented audit does not include surveillance or inspection for the purposes of process control or product acceptance.

EFFECTIVE DATE OF CONTRACT–The date of commitment to a contractor or the date on which the contractor began performance of work, whichever is earlier. The effective date of the contract is nor necessarily the date on which the contract is signed.

ESCALATION–This is a clause included in a contract when inflation is unpredictable. Its purpose is to protect the contractor against loss due to rising prices.

EXPERT–Where a dispute arises between the contracting parties which cannot be resolved by mutual agreement, some contracts will make provision for an expert to be appointed by both parties as a preliminary step before arbitration. The expert, whose qualifications are somewhat similar to those of the arbitrators, will study the disputed matter and hand down his ruling. His judgment is not necessarily binding but if the disputants can agree with it, a considerable amount of time and money is saved.

EXTRA WORK–Any work not contemplated by the contract. Always a scope change; always costs money and always results in a contract variation order.

FIXED RATES–Arrangements to pay contractor for certain costs which it will incur in connection with work under a cost reimbursable contract such as overheads but not including profit.

FORCE MAJEURE–An unexpected and disruptive event or circumstance which may operate to excuse a party for delayed performance under a contract. If a party is delayed in the performance of a contract and such a delay is caused by acts of God, war, riots, civil insurrection, acts of the public enemy, strikes, lock-outs, accidents, acts of civil or military authority, fires, floods, earthquakes, or windstorms beyond the reasonable control of the party delayed, such delay shall be added to the time for performance of the obligation delayed, unless the date, schedule, or time period for the performance of the obligation is expressly stated in the contract to be guaranteed. The contractor is not normally entitled to additional or extra compensation by reason of having been delayed but may have the period of delay added to the contract completion date or nearest milestone date. Of course, if the client wishes to buy back this period in order to finish on the original schedule, he may do so.

FREE ISSUE MATERIAL–This is construction material (or materiel) issued to the contractor by the client for the work, usually because the client's organization or that of the managing contractor has a more effective procurement, inspection, and expediting facility than the construction contractor. In many areas, the oil company client has a duty-free arrangement with the Government which may not be available to the contractor.

FREE ON BOARD (FOB)–The type of shipping contract in which the buyer provides the ship and the seller provides the cargo at port of loading.

HOLD POINT–An agreed quality control inspection stage beyond which work may not proceed except by written agreement from the inspector.

INDEMNITY AND INSURANCE–Before work may begin, the contractor is expected to provide proof of insurance against certain risks. The client may indemnify the contractor against certain other risks that are not recoverable under insurance policies held by the contractor (see Chapter 6).

INFORMATION TO BIDDERS–A set of standard and specific instructions to bidders together with data that defines the basis for bid preparation and bid submission.

INFORMATION FOR PROPOSAL–The document package issued to solicit proposals by contractors. The package consists of the following:

(a) Information to bidders
(b) Pro forma contract
(c) Proposal form

JOB EXPLANATION MEETING–A meeting arranged by the client to explain to all bidders the nature of the work involved in the Request for Quotation. Ideally, the job explanation meeting should follow the distribution of the bid packages so that bidders may have the opportunity to read through the package and prepare questions to raise at the meeting. It is usual for the job explanation meeting to take place on the same day as a site visit to the future work area, but in a large contract this is not always practical.

KICKOFF MEETING–See PRE-CONSTRUCTION MEETING.

LETTER OF INTENT–Advice to the successful bidder in the form of a letter, fax, or telex, the text of which will be similar to the following:

> This letter (fax/telex) confirms our intent to award a contract to John Smith Inc. for the installation of a 24" pipeline from the Westside Refinery to the Eastside Terminal together with associated construction work.
>
> The Scope of Work shall include that work described in the documents made part of Inquiry No. 123456
>
> Pricing for the work shall be as described in Part IV of the Inquiry and shall total $10,000,000.
>
> Please confirm by return letter (fax/telex) that you are in agreement with the above provisions.
>
> The effective date of the contract shall be the date we receive your confirmation by letter (fax/telex).

If there were any addenda to the scope during the bid period, these would be mentioned in the letter of intent together with prices quoted. The purpose of the letter of intent is to enable the contractor to start to mobilize and generally prepare for the work before the contract is signed.

LIQUIDATED DAMAGES–Compensation for late completion or walkout by a party to the contract. In certain civil contracts compensation May be stipulated as a daily or weekly sum in the bid but in most oil industry related contracts the amount is mentioned as a percentage of the contract price for each day the work is delayed by the contractor. Such contracts are generally silent on a reverse situation.

MANAGING (OR MANAGEMENT) CONTRACTOR–A contractor engaged by a client to assist in managing a project.

MECHANICAL COMPLETION–The stage reached when the permanent works described in the contract scope of work, or a portion or portions thereof, have been mechanically and structurally put in a tight and clean condition and otherwise constructed as provided in the contract. All deficiencies, including those which could prevent or delay safe and orderly prestart-up/start-up procedures by the client or timely achievement of permanent works operation at the conditions specified in the contract (to the extent that said deficiencies can be determined) must have been corrected. The client may, for the purposes of bonus payments to the contractor or by agreement with the contractor, consider that substantial completion has been reached when some minor work remains to be done (See SUBSTANTIAL COMPLETION).

MILESTONES, MILESTONE PAYMENTS–Intervals along the critical path of the contractor's master construction schedule at which completion should be reached of a major and usually vital portion of the work. In certain contracts, payment will be made not on percentage of progress per month but on achievement of the next milestone.

NOTICE OF ACCEPTANCE–Notice given by client to a contractor that the work performed by said contractor is accepted, either for purposes of final payment or for purposes of final acceptance of the work under a given contract.

NOTICE OF COMPLETION–Notice given to the client by the contractor that he has completed performance of the work.

NOTICE OF REJECTION–Client's reply to a contractor's notice of completion specifying corrective or incomplete work remaining prior to issuance of client's notice of acceptance.

OPEN BIDDING–A method of contract procurement in which all financially and technically qualified contractors may submit a bid for a publicly advertised contract.

PRECONSTRUCTION MEETING–In theory, if not in practice, the successful bidder has dealt mainly with the client's contract personnel throughout the bid period and up to the signing of the contract. One of the reasons for holding the preconstruction meeting (otherwise known as the kickoff meeting) is to introduce the successful bidder (known as the contractor) to the construction team who will be administering and monitoring the progress of the contract work. This formality over, the client's team will get down to the business of discussing with the contractor his needs and his intentions during the start-up period. For example, although the contractor may be aware that he is to be supplied with water and electricity, because the contract says so, he may not know exactly where these sources may be. The site team will want to know what drawings the contractor requires at the beginning of the work and will also discuss with him the handling of his progress payments—most important to the contractor!

PRIME CONTRACT–A contract between client and a managing contractor assigning the latter overall responsibility for the project.

PROGRESS INVOICE–A formal invoice for payment of identifiable material, labor, or services mutually agreed as having been supplied or performed in progress for partial payment of the contracted work.

PROGRESS MEETING–A meeting held onsite at regular intervals, usually weekly, between the client's site construction team and the contractor's site team. The purpose of the meeting is to report progress and to discuss and record matters arising from construction and related activities. The minutes of these meetings could be of great importance in future claims discussions and for this reason, if none other, it is essential that the accuracy of the minutes is agreed upon by both parties, either by signing off the minutes after the meeting or by approving the minutes at the next meeting. The question of who should actually record the minutes of a progress meeting (client or contractor) has been debated in the past but it could be considered that the side who writes the minutes has the edge.

PROGRESS PAYMENT–Payment for work as it is completed by measurement of work as it is in place and applying the pricing basis as specified in the contract.

PROJECT PROCEDURES–A set of guidelines and instructions approved and issued by the client's project management to each client discipline and department engaged on the project and describing the methods to be adopted for the operation of the respective disciplines and departments.

PROJECT PROPOSAL–A document containing process and design information for the proposed facilities including requirements applicable to mechanical features and construction materials. The project proposal may be compiled by the client or by a contractor who is engineering the proposed facilities. Should this contractor be required to act as managing contractor, the proposal may contain a certain modus operandi but should not be confused with a request for quotation.

PROJECT RECORD BOOK–A collection of drawings and data compiled by the contractor to include all important items of equipment installed on the project. The PRB is valuable as a reference during the initial start-up of the plant and also during subsequent operation, maintenance, and inspection. It provides information necessary for future checks of equipment performance and possibly for planning future plant extension or plant design. The project record book will often be a required supplement to the contractor's as built documentation.

PUNCH LIST–A listing of work items remaining unsatisfactory or incomplete that must be performed satisfactorily by the contractor prior to the issue of a final Notice of Acceptance.

QUALITY ASSURANCE PROGRAM–A comprehensive system of plans, procedures, instructions, and documentation describing the measures necessary to ensure that construction quality is planned and obtained. The Q.A. Program will consist of descriptions and procedures for all aspects of execution of the contract's work, including work management, work planning, inspection of vendor supplied material, fabrication instructions, material control, quality control (inspection and testing), safety, and documentation. The contractor is usually required to produce a Q.A. manual containing a detailed program before work commences.

QUALITY CONTROL–The implementation of the Q.A. Program.

QUALITY CONTROL INSPECTION–The part of the quality assurance that by measurements, tests, or investigation, determines whether the product or service is in accordance with the prescribed quality requirements.

QUANTITY SURVEYOR–This is a British job title having its origin in the building industry and meaning one who measures quantities (of building materials) and assesses their value. In Europe recently, the QS has been used extensively in the oil industry, particularly in North Sea operations. His professional value has been recognized, not as a measurer or number cruncher but as a skilled estimator in change proposals and as the contractor's right-hand man in the claims department.

REIMBURSABLE CONTRACT–A contract whose payment structure provides for the direct reimbursement to a contractor of all costs incurred by him in performing a particular item of contracted work plus agreed upon markups. This type of pricing structure is frequently referred to as a cost-plus pricing formula. The contractor's markup or profit margin may be in the form of a fixed fee or a percentage of the whole cost of the work. It is usually to the client's advantage to obtain a fixed fee quotation from the contractor. The reimbursable type of contract is not the most favorable from the client's viewpoint but in the absence of engineering work prepared before the bid period, it is sometimes a necessary evil.

REIMBURSABLE COSTS–Costs for performing the work under a reimbursable cost contract for which the contractor receives *direct* reimbursement as opposed to *indirect* reimbursement via fixed rates.

REMEDIAL ACTION–Action by the contractor to correct nonconformance, error, or violation of procedures.

REPAIR–The process of restoring a nonconforming characteristic to a condition such that the item's capability to function reliably and safely is unimpaired, even thought that item still may not conform to the original requirement.

REQUEST FOR QUOTATION (BID PACKAGE) The document by which the client requests various contractors to submit proposals on given items of contracted work.

RETENTION–An amount of money withheld from payment to a contractor to assure completion of the contract. This money is usually based on a percentage of each payment and is released upon final acceptance of the work. A bank guarantee may be accepted in lieu of retention.

RETURN ON ASSETS (ROA)–The net profit after tax expressed as a percentage of the total money invested in an enterprise.

RETURN ON INVESTMENT (ROI)–The net profit after tax expressed as a percentage of the total money invested in an enterprise.

REWORK–The process by which a nonconforming item is made to conform to a specified requirement by completion, remachining, reassembling, or other corrective means.

RISK CAPITAL–Equity capital raised to finance a development that has technical, economic and other risks attached to it and thus cannot guarantee a return on the investment.

SCHEDULE FLOAT TIME–This is slack time available on the contract master schedule. Activities containing float time do not lie on the critical path.

SCOPE OF WORK– A description of the work to be performed by a contractor.

SELECTIVE BIDDING–A method of contract procurement in which only a limited number of contractors considered technically and financially qualified to perform the proposed work are invited by letter to submit bids.

SHORT FORM CONTRACT–This document has an abbreviated format compared to that of a standard contract and is used for lump sum or unit rate contracting of limited value and time. The SFC is designed for prompt execution by contracts staff and, provided the requisite number of bidders are obtained, contract award may be made in hours rather than in weeks. The SFC is largely in pre-printed format with spaces for the scope of work, usual terms and conditions, compensation, and special terms, if any. Because of the limited value of the SFC, usually $50,000 or less, the level of signature authorities is not so high and may probably be executed at site level without senior management review. The nature of the SFC precludes progress payments and settlement is made after completion of the work. Change orders are not used in the administration of the SFC.

SITE INSTRUCTION–A written instruction from the client's site team to the contractor to direct action within the contract's general scope of work. It is not a variation order and should entail no additional cost or time to perform. It instructs the contractor to do something or not to do something at short notice and in quick reaction situations. It may be used to alter or discontinue a practice that may be or may result in safety violation. If there is no real urgency about the communication, the despatch of a letter is preferable since the site instruction is usually in pre-printed format.

SITE REPRESENTATIVE–A supervisor appointed to act on behalf of the client's company representative in the field to monitor, review, and provide guidance in the contractor's performance of the work.

STANDBY TIME–The period (s) in which the contractor's equipment or personnel are committed exclusively to the contract work and are ready for immediate use in the work but cannot perform the work due to circumstances within the sole control of the client.

SUBSTANTIAL COMPLETION–A stage before mechanical completion when the client will accept the work carried out thus far because circumstances prevent the contractor from proceeding further.

SUPERPROJECT–A major construction venture valued in excess of an arbitrary figure of $1 billion.

TURNKEY CONTRACT–Hard money contract in which contractor is responsible for the complete installation of the plant or facilities ready for client occupancy and operation.

UNIT PRICE–A firm price that is applied to a measurable unit of work and/or quantity of material.

UNIT RATE–The applicable charge for a given unit of time, usually an hour, involving the use of contractor's labor and/or equipment.

Unit rates may be used as the reimbursable cost element in a contract or as the basis for pricing changes and for settling standby claims of a contractor.

VERIFICATION–Investigation to confirm that an activity, product or services conforms to specified requirements. Verification differs from inspection in that verification confirms that a prescribed inspection has been carried out.

WORK BREAKDOWN STRUCTURE (WBS)–A "family tree" type of subdivision of the complete project work scope. The WBS identifies the major items needed to accomplish the project objectives and provides the framework for the establishment of project baselines for cost, time, and resource.

WORK ORDER–A written order signed by authorized client representatives directing the contractor to make changes in the specifications, drawings, work schedule, quantity, and/or scope of work as allowed within the contract. Rules regarding the use of work orders will vary from client to client but a number of procedures will allow the work order to be used only where applicable unit rates and prices are already in the contract. In other words, only in cases precluding separate negotiation.

WORK PACKAGE (WP)–A defined task which is identified in the WBS as a level of work where cost, schedule, and resources may be expressed. Work packages will be defined in the contract's scope of work and in the compensation section. Clients working with this system will require the contractor to divide his monthly progress invoices into work package activities regardless of the sequence of scheduled activities.

WORK PROCEDURE (WORK INSTRUCTIONS, JOB INSTRUCTIONS, etc.)–After contract award but before work starts the contractor is usually given a period of time to prepare, for client approval, the modus operandi for each stage of the work. The standard of the paperwork required varies but as a general comment it might be said that where the highest standards of safety and precision are required (i.e., offshore work and accommodation structures), the requirements of the work procedures are that much more stringent.

WAITING ON WEATHER (WOW)–Standby time due to inclement weather, usually in offshore work.

Index